Bernhard Zachhuber

Explosives Detection via Stand-off Raman Spectroscopy

AF113989

Bernhard Zachhuber

Explosives Detection via Stand-off Raman Spectroscopy

Investigation from a safe distance

Südwestdeutscher Verlag für Hochschulschriften

Impressum/Imprint (nur für Deutschland/only for Germany)
Bibliografische Information der Deutschen Nationalbibliothek: Die Deutsche Nationalbibliothek verzeichnet diese Publikation in der Deutschen Nationalbibliografie; detaillierte bibliografische Daten sind im Internet über http://dnb.d-nb.de abrufbar.
Alle in diesem Buch genannten Marken und Produktnamen unterliegen warenzeichen-, marken- oder patentrechtlichem Schutz bzw. sind Warenzeichen oder eingetragene Warenzeichen der jeweiligen Inhaber. Die Wiedergabe von Marken, Produktnamen, Gebrauchsnamen, Handelsnamen, Warenbezeichnungen u.s.w. in diesem Werk berechtigt auch ohne besondere Kennzeichnung nicht zu der Annahme, dass solche Namen im Sinne der Warenzeichen- und Markenschutzgesetzgebung als frei zu betrachten wären und daher von jedermann benutzt werden dürften.

Coverbild: www.ingimage.com

Verlag: Südwestdeutscher Verlag für Hochschulschriften GmbH & Co. KG
Heinrich-Böcking-Str. 6-8, 66121 Saarbrücken, Deutschland
Telefon +49 681 37 20 271-1, Telefax +49 681 37 20 271-0
Email: info@svh-verlag.de

Approved by: Wien, TU, Diss., 2012

Herstellung in Deutschland (siehe letzte Seite)
ISBN: 978-3-8381-3259-4

Imprint (only for USA, GB)
Bibliographic information published by the Deutsche Nationalbibliothek: The Deutsche Nationalbibliothek lists this publication in the Deutsche Nationalbibliografie; detailed bibliographic data are available in the Internet at http://dnb.d-nb.de.
Any brand names and product names mentioned in this book are subject to trademark, brand or patent protection and are trademarks or registered trademarks of their respective holders. The use of brand names, product names, common names, trade names, product descriptions etc. even without a particular marking in this works is in no way to be construed to mean that such names may be regarded as unrestricted in respect of trademark and brand protection legislation and could thus be used by anyone.

Cover image: www.ingimage.com

Publisher: Südwestdeutscher Verlag für Hochschulschriften GmbH & Co. KG
Heinrich-Böcking-Str. 6-8, 66121 Saarbrücken, Germany
Phone +49 681 37 20 271-1, Fax +49 681 37 20 271-0
Email: info@svh-verlag.de

Printed in the U.S.A.
Printed in the U.K. by (see last page)
ISBN: 978-3-8381-3259-4

Copyright © 2012 by the author and Südwestdeutscher Verlag für Hochschulschriften GmbH & Co. KG and licensors
All rights reserved. Saarbrücken 2012

Abstract

Due to the worldwide threat of terrorism, a transportable prototype is being developed combining three optical technologies for the detection and identification of explosives (OPTIX) at a distance of 20 metres. The techniques are laser induced breakdown spectroscopy, pulsed laser fragmentation mid-infrared spectroscopy and Raman spectroscopy.

As part of this FP7 project, a stand-off Raman system was set up and optimised. The system consists of a pulsed laser (Nd:YAG 532, 355 or 266 nm) for sample excitation, a telescope for signal collection, a spectrograph to separate spectral information and an intensified charge-coupled device camera which is synchronised with the pulsed laser to avoid the collection of daylight and fluorescence.

To meet the challenging situations in real life scenarios, different explosives as well as their precursors were analysed in the presence of interfering substances and on a variety of background materials, such as plastics or different car doors. Furthermore, detection limits were established, showing that this technique is not limited to bulk detection. The influence of the excitation wavelength was investigated by changing the laser from visible light (532 nm) to ultra violet radiation (355 or 266 nm), which can increase the Raman signal significantly.

In addition to the OPTIX project requirements, the stand-off distance was successfully extended to 100 metres at a testing ground at the Austrian Armed Forces. Moreover, combining stand-off technology with spatial offset Raman scattering it is now possible to identify and quantify substances at up to 40 metres in bottles which are non-transparent to the human eye. This method permits the detection of concealed substances in "real life" containers from a safe distance, even if the container exhibits interfering fluorescence. An alternative way to probe substances inside turbid bottles was found by using depth-resolved stand-off Raman spectroscopy. This methodology employs the speed of light to distinguish between samples located at different distances from the setup.

Acknowledgement

I want to say THANK YOU to:

Bernhard Lendl
  for the possibility to work in the international environment of his research group
  for the chance to broaden my horizon on many conferences all around the world
  for the exciting research topic and the correction of my thesis

Engelene t. H. Chrysostom
  for her enthusiasm when sharing her knowledge
  for the technical assistance when the equipment needed nursing
  and for the correction of my thesis

Alison J. Hobro
  for the successful cooperation

Christoph Gasser and Georg Ramer
  for their endurance in the laboratory

my colleagues for inspiring discussions about science and real life topics
  A. Genner, C. Carrillo-Carrión, C. Reidl-Leuthner, C. Wagner, C. Koch
  E. González García, E. Aguilera-Herrador, H. Moser, J. Frank, J. Kuligowsky,
  K. Wieland M. C. Alcudia León, M. Brandstetter, S. Radel, T. Furch, W. Ritter
  W. Tomischko

colleagues from other institutions for their unbureaucratic cooperation
  A. Wieser, E. Buchebner, P. Zachhuber (Institute of Geodesy and Geophysics)
  M. Fuchs, T. Mikats (Inst. for Production Engineering and Laser Technology)
  K. Föttinger (Institute of Materials Chemistry)
  H. Oppenheim, B. Schneider and co-workers (Austrian Armed Forces)

the numerous OPTIX partners
  for their collective effort to bring the prototype to life
  special thanks to the colleagues at the Swedish FOI for sharing their knowledge

Mama und Papa
  für ihre Unterstützung in jeder Hinsicht

my friends
  simply for their friendship and the numerous, incredible volleyball matches

Since money makes the world go round, I also want to thank
  the European Community's Seventh Framework Program (Grant No. 218037)
  the Defence Science and Technology Laboratory, trading fund of the UK MOD
  the FFG, Research Studios Austria program

CONTENT

ABSTRACT I
ACKNOWLEDGEMENT II
LIST OF PUBLICATIONS V
LIST OF ABBREVIATIONS VI

1 INTRODUCTION 1

1.1 Statistical information on terror 1
1.2 OPTIX: Optical Technologies for the Detection and Identification of Explosives 2
1.3 Stand-off Raman spectroscopy 7
1.4 Detection of explosives 8
1.5 Commercial products for spectroscopic detection of explosives 15
1.6 Investigated Substances: Raman reference spectra and properties 18

2 SETUP AND CHARACTERISATION OF STAND-OFF SYSTEM 26

2.1 Setup overview 26
2.2 Setup details 28
2.3 Influence of laser power and measurement time on signal quality 52

3 APPLICATIONS OF STAND-OFF RAMAN SPECTROSCOPY 62

3.1 Qualitative analysis 62
3.2 Quantitative analysis 68
3.3 Stand-off Raman analysis via UV telescope 70

4	**STAND-OFF SPATIAL OFFSET RAMAN SCATTERING**	**77**
4.1	Stand-off SORS principle	77
4.2	Line scan	78
4.3	Beam broadening with plastic thickness	82
4.4	SORS in real world containers	86
5	**DEPTH-RESOLVED STAND-OFF RAMAN SPECTROSCOPY**	**93**
5.1	Measurement principle	93
5.2	Application of depth-resolved stand-off Raman spectroscopy	93
6	**FUTURE PLANS FOR STAND-OFF RAMAN SPECTROSCOPY**	**96**
6.1	Three-dimensional stand-off Raman imaging	96
6.2	Stand-off SORS imaging	96
7	**REFERENCES**	**102**
8	**PUBLICATIONS**	**113**
8.1	Publication 1	111
8.2	Publication 2	133
8.3	Publication 3	145
8.4	Publication 4	155
8.5	Publication 5	165
8.6	Publication 6	171
8.7	Publication 7	177
8.8	Publication 8	181

List of Publications

1.) "**Standoff trace detection of explosives – a review**"
S. Wallin, A. Pettersson, H. Östmark, A.J. Hobro, B. Zachhuber, B. Lendl, M. Mordmüller, C. Bauer, W. Schade, U. Willer, J. Laserna and P. Lucena; Proceedings of 12th Seminar on New Trends in Research of Energetic Materials (2009) ISBN 978-80-7395-156, 349-368

2.) "**Stand-off Raman Spectroscopy of Explosives**"
B. Zachhuber, G. Ramer, A. Hobro and B. Lendl; Proc. of SPIE Vol. 7838 (2010)

3.) "**Stand-off Raman spectroscopy: a powerful technique for qualitative and quantitative analysis of inorganic and organic compounds including explosives**"
B. Zachhuber, G. Ramer, A.J. Hobro, E. t. H. Chrysostom and B. Lendl; Anal. Bioanal. Chem. 400 (2011)

4.) "**Stand-off spatial offset Raman spectroscopy – A distant look behind the scenes**"
B. Zachhuber, C. Gasser, A. Hobro, E. t. H. Chrysostom, B. Lendl; Proc. of SPIE Vol. 8189 (2011)

5.) "**Stand-Off Spatial Offset Raman Spectroscopy for the Detection of Concealed Content in Distant Objects**"
B. Zachhuber, C. Gasser, E. t. H. Chrysostom and B. Lendl; Analytical chemistry 83 (2011)

6.) "**Quantification of DNT isomers by capillary liquid chromatography using at-line SERS detection or multivariate analysis of SERS spectra of DNT isomer mixtures**"
B. Zachhuber, C. Carrillo-Carrión, B.M. Simonet Suau, B. Lendl; Journal of Raman Spectroscopy, accepted

7.) "**Spatial Offset Stand Off Raman Scattering**"
B. Zachhuber, C. Gasser, E. t. H. Chrysostom, B. Lendl; in Lasers, Sources, and Related Photonic Devices, OSA Technical Digest (CD) (Optical Society of America, 2012), paper LT2B.4

8.) "**Depth profiling for the identification of unknown substances and concealed content at remote distances using time resolved stand-off Raman spectroscopy**"
B. Zachhuber, C. Gasser, G. Ramer, E. t. H. Chrysostom and B. Lendl; Applied Spectroscopy, accepted

List of Abbreviations

ANFO	Ammonium Nitrate Fuel Oil	LIDAR	Light Detection And Ranging
APCI	Atmospheric Pressure Chemical Ionisation	LIF	Laser Induced Fluorescence
ATR	Attenuated Total Reflection	LOD	Limit Of Detection
CARS	Coherent Anti-Stokes Raman Spectroscopy	MS	Mass Spectrometry
CCD	Charge Coupled Device	NCTC	National Counterterrorism Center
CRDS	Cavity Ring Down Spectroscopy	OPTIX	Optical technologies for the detection and identification of explosives
CWAS	Chemical War Agent Simulant	PE	Polyethylene
DIAL	Differential absorption LIDAR	PETN	pentaerythritol tetranitrate
DIMP	Diisopropyl methylphosphonate	PLF	Pulsed Laser Fragmentation
DKDP	Deuterated potassium dihyrogenphosphate	PMMA	Poly (methyl methacrylate)
DNT	Dinitrotoluene	PP	Polypropylene
EGDN	Ethylene glycol dinitrate	QCL	Quantum Cascade Laser
FOV	Field Of View	QEPAS	Quartz Enhanced Photo Acoustic Spectroscopy
FWHM	Full Width at Half Maximum	RDX	Research Department Explosive
GC	Gas Chromatography	RMS	Root Mean Square
HDPE	High density polyethylene	S/N	Signal-to-Noise ratio
HMDT	Hexamethylenetriperoxide diamine	SERS	Surface Enhanced Raman Scattering
HMX	Cyclotetramethylen-tetranitramine	SORS	Spatial Offset Raman Spectroscopy
HPLC	High Pressure Liquid Chromatography	SPME	Solid Phase Micro Extraction
ICCD	Intensified CCD	SWIR	Short wave infrared spectroscopy
IED	Improvised Explosive Devices	TATP	Triacetone triperoxide
IMS	Ion Mobility Spectroscopy	TEPS	Townsend Effect Plasma Spectroscopy
IR	Infrared	TNT	Trinitrotoluene
LDPE	Low density polyethylene	VIS	Visible
LIBS	Laser Induced Breakdown Spectroscopy	YAG	Yttrium Aluminium Garnet

1 Introduction

1.1 Statistical information on terror

A major priority of most governments around the world is the security and safety of its citizens. According to the National Counterterrorism Center (NCTC), in the year 2010 alone, more than 11500 terrorist attacks were committed. This was an increase of 5% over the previous year. These attacks resulted in approximately 50000 victims, including 13200 deaths. The nature of these attacks between 2005 and 2010 are shown in Fig. 1a.

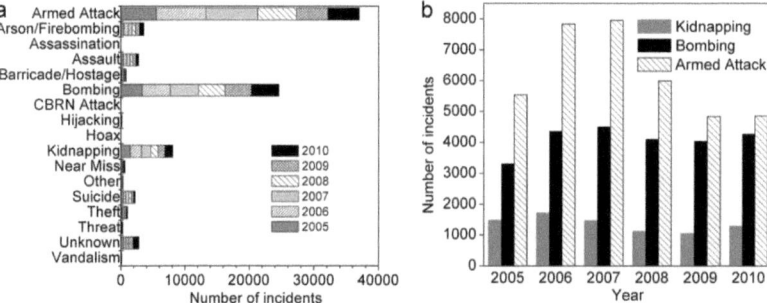

Fig. 1: Terror attacks between 2005 and 2010; a: types of terrorism; numbers for individual years are marked by differently shaded bars; b: change over time; armed attacks, bombing and kidnapping account for more than 80% of total incidents; the National Counterterrorism Center (NCTC) provided the presented statistical data which was published in August 2011 by the United State Department of State in "Country Reports on Terrorism 2010"

In 2010 more than a third of the terror incidents were armed attacks, followed by bombing and kidnapping as shown in Fig. 1b. Whereas the number of armed attacks decreased during the past three years, the number of bombings remained significant. This is particularly alarming since nearly 70% of all deaths are caused by bombings, including suicide attacks. In order to reduce the number of terror incidents, as well as their severity, it becomes essential to unambiguously identify explosives and potentially harmful substances. In order to meet in part these demands, this work was executed as part of the OPTIX EU FP7 project which aimed at developing technologies for the detection and identification of Explosives. A brief description of the project is outlined below.

1.2 OPTIX: Optical Technologies for the Detection and Identification of Explosives

Within the EU FP7 project OPTIX (Optical Technologies for the Detection and Identification of Explosives) a transportable prototype for the detection of suspicious substances from a distance of 20 m is being developed. In this project three laser based analytical techniques are combined to allow improved discrimination between explosives and harmless materials: Raman spectroscopy, PLF-IR (Pulsed Laser Fragmentation-Infrared Spectroscopy) and LIBS (Laser Induced Breakdown Spectroscopy), as illustrated in Fig. 2. For the OPTIX prototype the same central laser is used consecutively for each technique. The individual requirements concerning the laser energy density (Fig. 2) are achieved by changing the laser spot size on the sample, via an adjustable beam expander.

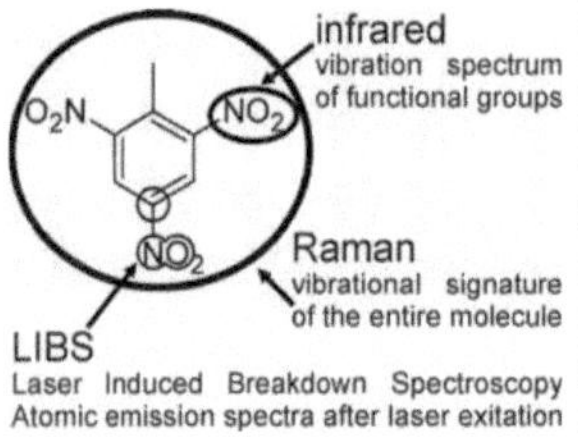

		Raman [1]	PLF-IR [2]	LIBS [1]
	laser pulse energy (mJ)	108	87	104-568
	laser spot size (mm^2)	47	12.5	2 - 8
	energy density (mJ/mm^2)	2.3	7.0	52 - 71

Fig. 2: TNT molecule and beam parameters for individual techniques; infrared spectroscopy specifically investigates functional groups, Raman spectroscopy analyses the entire vibrational structure, whereas LIBS studies the atomic composition of a molecule

The non-destructive Raman spectroscopy technique investigates the whole molecular structure via the vibrational behaviour of the analyte. With IR spectroscopy after sample material is evaporated by a fragmentation laser, the formed plume is analysed for NO and NO$_2$ via two QCLs (Quantum Cascade Lasers). NO and NO$_2$ are substance fragments particularly interesting for the detection of explosives containing nitro-groups. LIBS yields the atomic composition of the probed material. A laser ignites a plasma at the sample surface and the atomic emission lines allow the determination of the composition of the probed material. The combined information of all three subsystems is expected to increase the confidence with which the discrimination between explosives and harmless substances is made possible. To avoid human intuition, an automated chemometric algorithm will analyse the acquired data. A brief introduction of the three individual techniques used in the OPTIX project is given below.

1.2.1 Raman Spectroscopy for OPTIX subsystem

Raman spectroscopy measures vibrational transitions in a molecule or sample through the analysis of inelastically scattered light and, as this information is specific to the particular chemical structure in a molecule, Raman spectra can be thought of as a fingerprint of a sample. Raman spectroscopy usually employs a laser in the visible, near ultraviolet or near infrared range. With this non-destructive technique a wide range of substances in gaseous, liquid and solid forms can be analysed. Fig. 3 illustrates the principle of stand-off Raman spectroscopy.

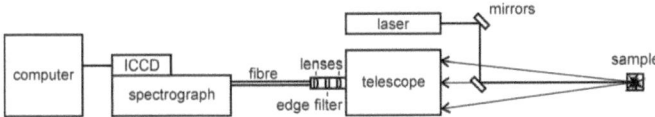

Fig. 3: Stand-off Raman system; a laser excites the sample and the telescope collects the Raman scattered light; the light is passed through an edge filter, to remove Rayleigh scattered light, and then guided into a spectrograph via a fibre optic cable; the intensified CCD camera (ICCD) allows gated detection reducing contribution from fluorescence and ambient light

Raman spectroscopy was used to study nitrogen as well as peroxide based explosives in bulk and trace quantities. Explosives on various surfaces, such as metals or plastics, and in different containers, such as glass bottles can be detected. Using pulsed laser systems Raman spectra were recorded under challenging weather conditions including fog and rain at remote distances ranging from 3 to 533 m [3]. Fig. 4 shows a Raman spectrum of TNT (trinitrotoluene) measured at a distance of 100 m.

INTRODUCTION

To obtain good quality stand-off Raman spectra in daylight, it is necessary to use a pulsed laser which is synchronised to the detector to collect the entire Raman signal and to avoid most of the disturbing daylight.

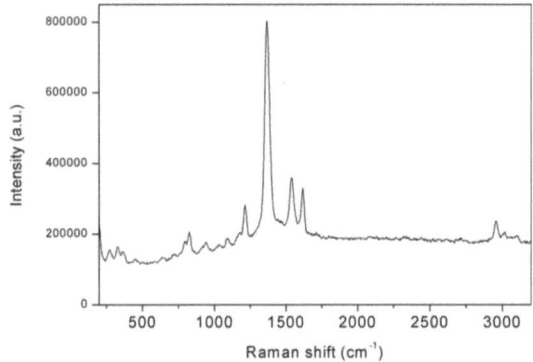

Fig. 4: 100 m-stand-off Raman spectrum of TNT (measured at the Austria Armed Forces)

1.2.2 Pulsed Laser Fragmentation – Mid-IR Spectroscopy for OPTIX subsystem

With infrared (IR) spectroscopy, characteristic absorption features typically between 3 and 20 µm allow the identification of substances. The use of Pulsed Laser Fragmentation (PLF) generates volatile products which are detected. The plume formed during this process is probed by two Quantum Cascade Lasers (QCLs) operating at different wavelengths to analyse specific absorption bands, e.g. 5.3 and 6.3 µm for probing NO and NO_2, respectively. The ratio of the two analyte concentrations is used to distinguish between different substances. In Fig. 5 the principle of stand-off PLF-IR is exemplified.

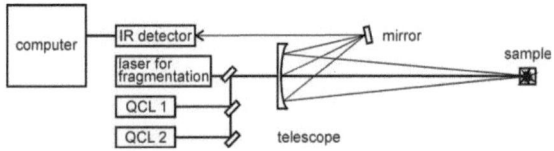

Fig. 5: Stand-off infrared system; sample molecules are fragmented and the generated plume is probed by two QCLs at different wavelengths

Stand-off IR analysis of explosives has concentrated on direct sensing of nitrogen based explosives such as TNT (trinitrotoluene) and HMX (cyclotetramethylen-tetranitramine) through the simultaneous detection of NO and NO_2 in the sample and calculating the ratio between the two (Fig. 6).

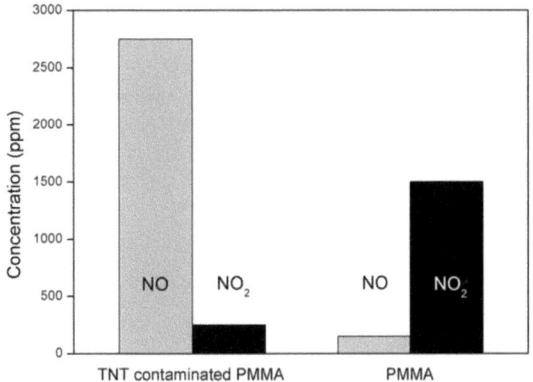

Fig. 6: Ratio of the NO/NO_2 production by PLF for TNT on a PMMA surface (modified from [4])

Studies have also been performed on peroxide based explosives such as TATP (triacetone triperoxide), relying on their relatively vapour pressure in conjunction with careful selection of the laser wavelength in order to record spectra using a small hand held probe. For TATP a QCL emitting between 1235.1 and 1245.3 cm^{-1} was used [4]. Current stand-off distances are on the order of 5 m but with improvements in collection optics it is envisaged that 20 m are possible.

1.2.3 Laser Induced Breakdown Spectroscopy – for OPTIX subsystem

Laser induced breakdown spectroscopy (LIBS) is based on atomic emission spectroscopy. A laser is focused onto a small area on the sample where it ablates a small amount of the sample surface creating a plasma. As the plasma cools there is a short time frame where characteristic atomic emission lines of the elements can be observed. LIBS is usually performed using a laser operating at 1064 or 532 nm and the full spectrum ranges from 200 to 980 nm. Although LIBS relies on ablation, the ablated area is so small that the technique can be considered as minimally destructive and the small ablation spot size allows for measurements with a high spatial resolution.

A sketch of a stand-off LIBS system is shown in Fig. 7.

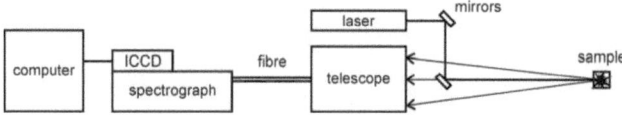

Fig. 7: Stand-off LIBS system; a high energy laser pulse generates a plasma at the sample surface; atomic emission lines generated by this event are collected by a telescope and guided into a spectrograph via a fibre optic cable; an intensified CCD camera records the spectrum and allows the reduction of white light background generated after plasma formation

Regarding explosives analysis, LIBS was used to study a range of different explosives, both nitrogen and peroxide based, at stand-off distances of up to 135 m [5].

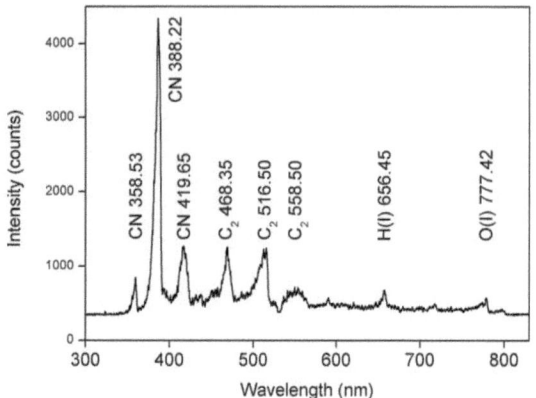

Fig. 8: 20 m-stand-off LIBS spectrum of TNT (provided by P. Lucena, University of Malaga, Spain)

1.2.4 OPTIX Prototype

The aim of the OPTIX project (www.fp7-optix.eu) is to combine the strengths of Raman spectroscopy, PLF-IR and LIBS for a robust identification of explosives. Due to the similar hardware requirements of all techniques (laser, optics, spectrometer) a prototype is been built on a single platform. The individual datasets of each spectroscopic method will be combined for chemometric analysis (University of Dortmund). Therefore, the result will be more specific and sensitive than the single techniques and it will be more difficult to confuse or defeat the system.

Experimental optimisation of the individual core technologies were carried out at the Vienna University of Technology (Raman), the Clausthal University of Technology (PLF-IR) and the University of Malaga (LIBS). The OPTIX prototype was assembled at

Indra in Spain. The central 532 nm-laser was developed by EKSPLA, Lithuania and the compact spectrograph was designed by Avantes, Netherlands. The prototype is been tested at the University of Malaga and further tests will be conducted at the Swedish Defence Research Agency (FOI), whose expertise is reflected by their numerous publications and their patent on "Stand-off Detection of hazardous Substances such as Explosives and Components of Explosives" (International publication number WO 2009/105009 A1).

1.3 Stand-off Raman spectroscopy

In 1974 the first stand-off Raman system was proposed [6]. Since then extraordinary advances have been made due to increased detector sensitivities, smaller laser systems and careful investigation and optimisation of different optical concepts. For example, the interference of ambient light for continuous detection can be circumvented by measuring in the solar blind spectral range (230-320 nm) [7] or by using a detector which is synchronised to a pulsed laser [8].

Numerous applications of stand-off Raman spectroscopy exist ranging from geology on earth [9] as well as for planetary exploration [10], [11] and archaeology [12] to chemical detection and explosive analysis [13], [3], [14]. For a more detailed view on stand-off Raman spectroscopy the reader is referred to review by Hobro et al. [15]. Furthermore, a review by Wallin et al. (attached at the end of this book) on "Laser-based stand-off detection of explosives" [16] illustrates the individual techniques in more detail, covering laser induced breakdown spectroscopy (LIBS), light detection and ranging (LIDAR), laser induced fluorescence (LIF), photo-fragmentation LIF (PF-LIF), pulsed laser fragmentation mid-infrared spectroscopy (PLF-IR), differential absorption LIDAR (DIAL) and coherent anti-Stokes Raman spectroscopy (CARS). There are different ways how to further improve stand-off Raman detection: better portability, method combination, longer measurement ranges, detection in adverse weather conditions, resonance Raman for lower detection limits, imaging capability or detection in non-transparent bottles as well as non-linear techniques such as CARS.

To improve the portability, a particularly small instrument was developed using a camera lens for a commercial SLR camera [17]. Others combined different laser based techniques such as LIBS and Raman spectroscopy to gain additional substance information [18] which can then be analysed applying a mathematical algorithm [19].

Stand-off Raman spectroscopy allows the identification of TATP through glass windows at 200 m or NH_4NO_3 at 470 m during heavy rainfall [20]. Furthermore, resonance Raman allowed detection of TNT, DNT and nitromethane vapour at the ppm level using a tunable UV laser [20]. Via a liquid crystal tunable filter, hyperspectral imaging of bulk material (sulfur, DNT, TNT and NH_4NO_3) was achieved at a distance of 10 m [20] and further improved to allow imaging of particles [21], [22]. Reaching out to even longer sampling distances, sulfur (a particularly good Raman scatterer) was measured from a distance of 1500 m during night time to avoid interference from ambient light [23]. Also the variety of stand-off Raman applications was extended. For example structural information of Arctic ice was gained from 120 m distance [24] and samples investigated under conditions relevant to Venus exploration [25]. Picosecond lasers were successfully used to separate instantaneous Raman signals from fluorescence which typically exhibits lifetimes of 2 ns or less [26]. Stand-off Spatial Offset Raman Spectroscopy (SORS) was used to measure substances in fluorescence bottles [27]. A technique which also allows the analysis of substances in containers which are non-transparent to the human eye [28]. An instrumentally challenging way to enhance the Raman response of a sample is to move towards non linear Raman techniques such as Coherent Anti-Stokes Raman Spectroscopy (CARS) which was used to detect traces of KNO_3, sulfur and urea from a distance of 12 m via shaped femtosecond pulses [29]. Furthermore, RDX was identified from a distance of 5 m within a time of 3 s [29]. In addition, this non-linear technique allows the detection of various explosives [30], [31], [32].

Alternative analytical techniques for the detection of explosives are introduced in the next section.

1.4 Detection of explosives

A variety of analytical methods is capable of explosive detection. This section briefly illustrates the numerous alternative technologies together with their properties. Due to the vast number of publications devoted to explosive detection only an overview can be given here.

Since this work is focused on the stand-off detection of substances, the parameter "distance" is of particular importance when discussing the individual techniques in the following. For the analysis of potentially dangerous substances it is advantageous to maximise the distance between setup and sample, reducing the risk for the operator. This benefit is not limited to the investigation of toxins and explosives but also serves investigators who analyse samples which are simply out of direct reach, such as icebergs [24] or geological samples on other planets [33]. In this summary, which is focused on the identification of harmful material, stand-off methods are discussed, followed by techniques which necessitate close interaction with the analyte.

1.4.1 Stand-off methods for the detection of explosives from a distance

Most analytical tools suitable for the analysis from a safe distance facilitate electromagnetic radiation to interact with the probed matter. Different spectral regions are used to specifically take advantage of the properties of the applied radiation.

1.4.1.1 Microwave imaging

The application of microwaves allows to distinguish between materials with different dielectric constants. With a 94 GHz-microwave (wavelength 3.2 mm) the detection of metals was possible from a distance of 50 m, even if the samples were hidden behind lightweight materials [34]. This concept can be used in a portal design to detect hidden material under clothing and tell if the material is made of metal or plastic [35], [36]. One drawback of microwave imaging is, however, the lack of chemical information. This limitation however can be overcome by using radiation with shorter wavelengths such as infrared radiation.

1.4.1.2 Mid-infrared stand-off detection

Infrared spectroscopy relies on the substance specific absorption of specific wavelengths. The identification of substances deposited on a surface is possible as the analyte reduces the reflection of certain infrared wavelengths according to its specific absorption. This passive method facilitates infrared radiation reflected from surfaces for detection. For example, at a distance of 60 m war agents on aluminium as well as on low-reflectivity surfaces were identified via passive radiometry [37]. The difference in radiation from uncontaminated and contaminated surfaces depends on the specific absorption of the contaminant, thus allowing the identification of a substance. This concept is not limited to solid samples; via passive multispectral infrared imaging, war agent clouds at a distance of 2.8 km distance were successfully detected [38]. The sensitivity of passive sensing can be increased using an infrared source to actively illuminate the sample, rather than to rely on passive radiation. Mid-IR spectra (700-1800 cm^{-1}) of chemicals (DNT, TNT, PETN, RDX) on surfaces (400 $\mu g/cm^2$) were analysed via open path Fourier transform infrared spectroscopy (FTIR) in passive mode at 30 m and at 60 m in active mode using a globar for sample illumination [39]. Rather than heating the sample with a broadband infrared source (i.e. globar), the wavelength of a tunable CO_2 laser can be used. When the laser wavelength coincides with an absorption band of the molecules, the sample is heated and the change in temperature permits substance recognition [40]. A soil sample contaminated with TNT was placed at a distance of 150 m from the analyser. The temperature increase was significantly higher when this sample was irradiated with a laser wavelength of 1065.3 nm (a TNT absorption band) compared to 1059.1 nm irradiation, where no absorption occurs for TNT.

Alternatively – for TNT detection, a high performance IR imager can be used to detect a surface contamination at a distance of 5 m [41]. To obtain a TNT-sensitive image, a tunable Quantum Cascade Laser (QCL) embedded in an external cavity was used to specifically illuminate the probed area with a wavelength coinciding with a TNT absorption band at 7.43 μm. The reduced IR intensity caused by the specific TNT absorption, allowed the localisation of the contaminated surface regions.

To obtain sample information it can be advantageous to focus on the detection of specific molecular structures rather than the entire substance, especially for analytes with low vapour pressure. Pulsed Laser Fragmentation (PLF) applies a laser which produces a plume of molecular sample fragments. For the detection of explosives, the mid-infrared absorption of the formed sample plume can be determined from a distance of 5 m via two QCLs, specific for NO and NO_2 (5.3 and 6.3 μm) [4].

1.4.1.3 Stand-off laser induced breakdown spectroscopy

Application of stronger, focused laser pulses leads to the partial destruction of sample material. Laser induced breakdown spectroscopy (LIBS) ignites a plasma of the sample and the emitted light reflects the atomic composition of the probed material. The advantageous combination of a 266 nm Nd:YAG and a 10.6 μm CO_2 laser (Townsend Effect Plasma Spectroscopy, TEPS) allowed the detection of TNT at a distance of 40 m [42]. Since an exhaustive review on stand-off LIBS is beyond the scope of this work the reader is referred to a review on LIBS for the detection of explosives [43].

1.4.1.4 Laser induced fluorescence spectroscopy

Laser Induced Fluorescence spectroscopy (LIF) records the fluorescence response of a sample after laser irradiation. Since LIF spectra of more complex molecules do not contain very specific information, a fragmentation prior to the analysis can improve the substance identification. Laser induced fluorescence (LIF) after pulsed laser dissociation at 248 nm allowed the detection of TNT in the gas phase from a distance of 2.5 m [44]. This measurement was conducted 1 cm above a heated surface (28-43°C) covered with TNT.

1.4.1.5 Stand-off Raman spectroscopy

Raman spectroscopy facilitates the inelastic scattering when incident laser photons hit sample molecules. The substance specific scattering allows to distinguish between different analytes. The topic of stand-off Raman spectroscopy is explained in section 1.3 including several applications for the detection of explosives from a distance.

1.4.2 Detection of explosives in close proximity

Here methods for the analysis of explosives are introduced which require close sample contact. First, optical methods will be discussed, followed by mass sensitive systems, separation techniques and sensor devices.

1.4.2.1 Optical techniques for the detection of explosives in close proximity

The latter section introduced a range of techniques which allow the detection of explosives from a distance. These methods used electromagnetic radiation to bridge the distance between measurement apparatus and sample. These optical techniques with stand-off capabilities have applications for the detection of explosives in close contact as well. Selected examples for the individual technique are given here.

For example, attenuated total reflection Fourier transform infrared spectroscopy was applied to particles of three different explosives in fingerprints [45]. Via LIBS hazardous microbiological material was identified [46] and with laser induced fluorescence (LIF) the fluorescence response after laser irradiation was used to analyse proteins, bacterial spores and bacterial cells [47]. Conventional Raman signals can be enhanced when the laser excitation wavelength resonates with a sample absorption band which is beneficial for the identification of explosives [48]. TNT, RDX and PETN were measured at a laser wavelength of 229 nm, leading to a 1000-fold signal enhancement.

Two-dimensional resonance Raman spectra of TNT, RDX and PETN were recorded [49]. This was achieved by exciting the samples consecutively with 40 different wavelengths between 210 and 280 nm. An alternative way to enhance Raman signals is Surface Enhanced Raman Scattering (SERS), where usually nano-particles of silver or gold enhance the signal on the order of several magnitudes. Numerous recipes exist to produce different SERS substrates. For example CdSe/ZnS quantum dots were synthesised [50] to identify DNT isomers after capillary-liquid chromatography [51]. Portable systems for the detection of chemical and biological agents were developed [52] and DNT vapour was detected by SERS [53]. Despite the SERS detection of 4(5'azobenzotriazole-3,5-dimethoxyphenylamine) from a distance of 5 m [54], SERS typically requires close sample contact.

With Cavity Ring Down Spectroscopy (CRDS) the intensity loss over time of a laser pulse trapped between two highly reflective mirrors is influenced differently by different substances, allowing their determination. For instance, in the spectral range between 7 and 8 µm trace explosive vapours of TATP, PETN and RDX were detected via mid-infrared CRDS using a broadly tunable optical parametric oscillator [55]. More examples for the detection of explosives via CRDS are summarised elsewhere [56], together with various other photonic devices for explosive detection, such as photoacoustic sensors.

The combination of mid-IR excitation with acoustic detection is the basis of optoacoustic spectroscopy. When a laser excites an absorption band of a substance, the material heats up inducing higher pressure due to the expansion of the analyte with temperature. With an appropriate intensity modulation of the exiting laser a microphone picks up the pressure fluctuation. Quartz Enhanced Photo Acoustic Spectroscopy (QEPAS) significantly enhances the sensitivity of the method by exchanging the microphone with a quartz tuning fork [57]. Bauer et al. [58] applied the principle of QEPAS to the detection of TATP vapour, yielding detection limits of 1 ppm at atmospheric pressure. A variation of photoacoustic spectroscopy was applied to surface adsorbed TNT, RDX and PETN at a distance of 20 m [59]. To achieve the reported sensitivity of 100 ng/cm^2, spectral features between 7.38-8.0 and 9.25-9.80 µm were recorded via two tunable QCLs. A UV lamp was added to this setup to compare the spectral features before and after UV radiation. At 8 m stand-off distance surface concentrations of 700 ng/cm^2 are clearly identified [60].

1.4.2.2 Mass spectrometry and Ion Mobility Spectrometry

Mass Spectrometry (MS) is applied to the detection of explosives in various different ways. Its capability to identify substances in a short time makes this technique an important tool for forensic science. Especially the low limits of detection allow successful detection of trace contaminations on surfaces. For example, Atmospheric Pressure Chemical Ionisation (APCI) [61] in combination with tandem MS allowed the detection of TNT directly from the skin of a hand at a concentration of 3 pg/l [62]. Furthermore, limits of detection of 90-130 ng/l for nitrobenzene and 320 ng/l for 2,4-DNT were reached using laser ionisation, a technique that leads to little fragmentation, improving the analyte identification [63]. MS is often combined with other separation techniques, which are described below. Such hyphenated techniques combine separation capabilities with the specificity of MS.

An alternative to the instrumentally challenging MS is Ion Mobility Spectroscopy (IMS). Due to the lower vacuum requirements IMS is instrumentally less demanding than a mass spectrometer. This is the reason why IMS is commonly used to detect explosive traces on luggage in airports. The drift time of different sample ions in a weak electrical field allows to distinguish between analytes. For example a limit of detection for triacetone triperoxide (TATP) in toluene of 187 µg/ml was achieved by IMS [64]. The development of IMS for the detection of explosives and explosive related compounds during the last three decades is illustrated in a review by Ewing et al. [65].

1.4.2.3 Separation techniques for the determination of explosives

Separation techniques employ the physical and chemical properties of substances to separate mixtures. In contrast to spectroscopic methods, close interaction between sample and apparatus is essential to achieve sufficient analyte separation.

A powerful tool for the analysis of explosives is High Pressure Liquid Chromatography (HPLC). Gaurav et al. [66] summarised numerous HPLC methods for the detection of explosives in various matrices. The most widespread detection technique for HPLC is UV absorption spectroscopy due to its simplicity and robustness. More specific information can be obtained by analysing the separated substances with different detection units. For example mid-infrared spectroscopy allowed the on-line determination of TATP (triacetone triperoxide) and HMDT (hexamethylenetriperoxide diamine) after their separation [67]. Coupling of HPLC with MS (Mass Spectrometry) gains additional information. The determination of 14 selected nitroaromatic compounds (explosives and their degradation products) via HPLC coupled to a tandem MS yielded limits of detection between 4 and 114 µg/l [68]. Gaurav et al. [69] summarises methods which further decrease HPLC detection limits to the sub ppb range via preconcentration steps such as SPME (Solid Phase Micro Extraction). Explosives can also be separated via Gas Chromatography (GC). An example is the analysis of 24 emulsion explosives in complex matrix via the coupling of GC and MS [70].

Many separation techniques exist which can be coupled to numerous detection systems to meet the challenging requirements the detection of hazardous materials in realistic environments exhibits. For example explosives and their degradation products in soil environments were analysed after separation with capillary electrophoresis [71]. A review on capillary and microchip electrophoretic analyses of explosives gives a broader overview [72].

1.4.2.4 Field tests and sensors for the detection of explosives

Situations where potentially dangerous substances have to be detected often require quick on-site results. The previously introduced laboratory based separation techniques often do not fulfil these requirements despite their excellent analytical capabilities. Small, sampling units built for harsh environments are preferred for such situations. For example, a rapid, simple field test for the identification of TATP (triacetone triperoxide) and HMDT (hexamethylenetriperoxide diamine) relies on the formation of hydrogen peroxide due to UV radiation in the presence of these analytes. In the presence of H_2O_2 horseradish peroxidise catalyses the formation of a green radical cation of 2,2A-azino-bis(3-ethylbenzothiazoline)-6-sulfonate (ABTS) [73]. An alternative sensor concept measures the reaction heat of the TATP decomposition on metal oxide catalysts [74]. The achieved detection limit in the range of parts per million, allows the determination of the relatively volatile TATP in the gas phase, avoiding direct contact to this sensitive substance. Similar detection limits for TATP were achieved using a sensing system based on a quartz micro balance [75]. To extend the range of detectable threats, multiple electrochemical potentiometric gas sensors were combined for the identification of

PETN, TNT and RDX [76]. Alternatively, especially designed receptor molecules in combination with a silicon microring resonator for the sensing of TNT in air are used [77]. The triphenylene-ketal based receptor molecules selectively interact with TNT in a reversible way, allowing specific detection of the target molecule. Also devices which sense substances in liquid phase exist. For example, a sensor head determines the fluorescence reduction of a polymer fibre, which correlates with the concentration of TNT in aqueous solution [78].

For the field measurement a key parameter for an instrument often is its portability together with a quick and reliable result. This is the reason why many commercial systems exist for the determination of explosives, illicit drugs or harmful chemicals. The following section introduces various commercially available products for the spectroscopic detection of explosives.

1.5 Commercial products for spectroscopic detection of explosives

The number of available instruments for the detection of explosives emphasises the importance of this field. Here, the market analysis was limited to portable instruments based on Raman scattering. There are some handheld Raman instruments commercially available which are compact and robust. However, most of them require close contact to the sample and do not work at stand-off distances. Here a list of commercial Raman instruments is introduced with the individual figures of merits.

1.5.1 Thermo Scientific

Compact handheld instruments for Raman (FirstDefender RM) as well as for infrared spectroscopy (TruDefender FTi) are available which operate under extreme conditions (vibration, temperature, thermal shock, sand and dust). These instruments are capable of identifying a growing number of chemicals, including explosives, chemical warfare agents, narcotics, toxic materials and their mixtures. While the infrared model contains an ATR (Attenuated Total Reflection) diamond it therefore requires direct contact with the sample, the Raman instrument is able to work at a distance of 17 mm, allowing the analysis of the content in transparent bottles. Tab. 1 summarises the figures of merit of both instruments.

INTRODUCTION

Tab. 1: Figures of merit of handheld Raman instrument (FirstDefender RM) and infrared instrument (TruDefender FTi) from Thermo Scientific; pictures from http://www.ahurascientific.com/chemical-explosives-id/products/index.php

	FirstDefender RM	TruDefender FTi
Working distance	17 mm	0
Spectral range	250-2875 cm^{-1}	650-4000 cm^{-1}
Resolution	7-10.5 cm^{-1}	4 cm^{-1}
Laser wavelength	785 nm	-
Laser power	0-250 mW	-
Laser spot size	0.14-1.8 mm	-
Weight	0.8 kg	1.54 kg

1.5.2 Ocean Optics

A handheld PinPoint Raman system with a working distance of 5 mm is available. Further details are listed in Tab. 2.

Tab. 2: Figures of merit of handheld Raman instrument (PinPointer) from Ocean Optics; photograph from http://www.oceanoptics.com/products/pinpointer.asp

	PinPointer
Working distance	5 mm
Spectral range	200-2400 cm^{-1}
Resolution	10 cm^{-1}
Laser wavelength	785 nm
Laser power	5-500 mW
Laser spot size	< 0.2 mm
Weight	1.4 g

1.5.3 Bay Spec

Xantus and 1stGuard are two portable Raman spectrometers which are designed for field measurements (Tab. 3).

Tab. 3: Figures of merit of handheld Raman instrument (xantus-1 and First Guard) from Bay Spec; photographs from http://www.bayspec.com/product_detail.php?a_id=4&p_id=57

	xantus-1		First Guard	
Working distance	-		-	
Spectral range	200-3000 cm^{-1}		200-3000 cm^{-1}	
Resolution	10-12 cm^{-1}		10-12 cm^{-1}	
Laser wavelength	532, 785, 1064 nm		532, 785, 1064 nm	
Laser power	0-500 mW		0-500 mW	
Laser spot size	-		-	
Weight	2.2 kg		1.8 kg	

1.5.4 DeltaNu

This company offer several hand-held devices similar to those listed from other companies above. However, DeltaNu also offer two stand-off Raman instruments, the ObserveR and the ObserveR LR. The ObserveR is hand held, operates from 0.3 to 3 m stand-off distance, using a continuous laser and comes with onboard libraries specifically for explosives, toxic industrial chemicals and materials. Other than most of the commercial instruments the ObserveR LR uses a pulsed laser which allows daylight operation. With a stand-off distance of 10 to 25 m this product is similar to the Raman subsystem in the OPTIX prototype. Figures of merit are shown in Tab. 4.

Tab. 4: Figures of merit of Raman instruments (ObserverR and ObserverR LR) from DeltaNu; photographs from: http://www.intevac.com/intevacphotonics/deltanu/deltanu-products/portable-raman-spectrometers/

	ObserverR		ObserverR LR	
Working distance	0.3-3 m		10-25 m	
Spectral range	300-2000 cm^{-1}		200-2200 cm^{-1}	
Resolution	20 cm^{-1}		15 cm^{-1}	
Laser wavelength	785 nm		1047 nm (pulsed)	
Laser power	120 mW		1500 mW	
Laser spot size	-		-	
Weight	2.3 kg		15.9 g	

1.5.5 ChemImage

ChemImage is developing a systems for explosive detection using hyperspectral stand-off imaging via Raman spectroscopy, short wave infrared spectroscopy (SWIR) and fluorescence spectroscopy. An example image obtained at 40 m stand-off distance is presented on their website.

http://www.chemimage.com/markets/threat-detection/explosive/index.aspx

1.6 Investigated Substances: Raman reference spectra and properties

Reference Raman spectra were obtained using a confocal Raman microscope (LabRAM, Horiba Jobin-Yvon/Dilor, Lille, France) for comparison with the stand-off Raman spectra. Raman scattering was excited by a HeNe laser at 632.8 nm and a laser power of 7 mW. The dispersive spectrometer was equipped with a grating of 600 lines/mm. The detector was a Peltier-cooled CCD detector (ISA, Edison, NJ, USA). The laser beam was focused manually on the sample by means of a ×20 microscope objective.

1.6.1 Ammonium nitrate and ANFO

The easy availability of ammonium nitrate (NH_4NO_3) as fertiliser makes this substance a common substance in Improvised Explosive Devices (IEDs). In Fig. 9 the Raman spectrum of NH_4NO_3 is depicted with the most dominant band at a Raman shift of 1053 cm^{-1}.

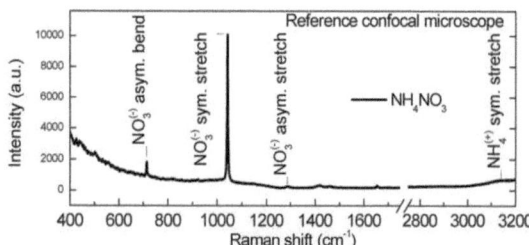

Fig. 9: Ammonium nitrate (NH_4NO_3) Raman reference spectrum (laser excitation 632.8 nm)

Furthermore, in combination with fuel oil (FO) ammonium nitrate (AN) is used for professional mining, usually as ANFO prills with a ratio of 94% AN and 6% FO. For terroristic acts ANFO mixtures are popular due to the easy availability of ammonium nitrate as fertiliser and diesel. In Fig. 10 a reference spectrum of ANFO is shown.

In comparison to a Raman spectrum of pure ammonium nitrate (Fig. 9) the baseline is elevated due to fluorescence caused by the fuel oil content of the sample.

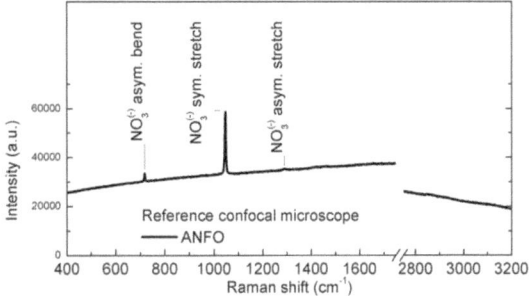

Fig. 10: Ammonium nitrate fuel oil (ANFO) Raman reference spectrum (laser excitation 632.8 nm)

The Raman bands present in the spectrum of ANFO (Fig. 10) are caused by vibrations of the nitrate group.

1.6.2 Ethylene glycol dinitrate (EGDN)

Ethylene glycol dinitrate (EGDN) is used in dynamite for colder weather to lower the freezing point of glycerine nitrate. In Fig. 11 a Raman reference spectrum of ethylene glycol dinitrate is illustrated.

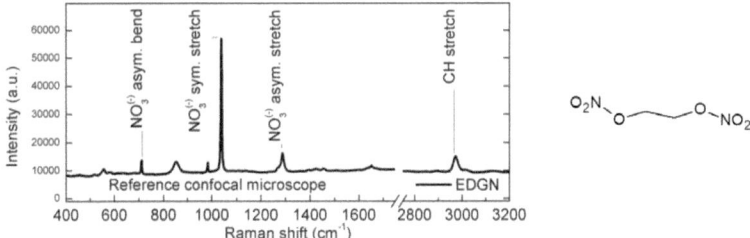

Fig. 11: Ethylene glycol dinitrate (EGDN) Raman reference spectrum (laser excitation 632.8 nm) and molecular structure

Since the investigated explosive did contain 61% NH_4NO_3 the according bands can be seen in Fig. 11.

1.6.3 Sodium chlorate and potassium chlorate

When analysing Improvised Explosive Devices (IED) chlorates are important substance due to their accessibility on the marked as weed killer. In Fig. 12 a Raman spectrum of sodium chlorate (NaClO$_3$) is shown measured on a confocal microscope.

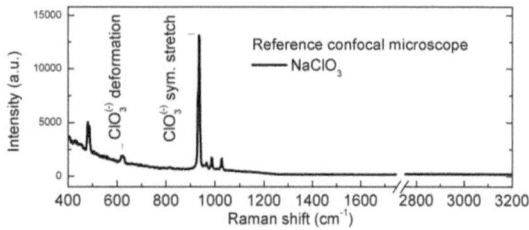

Fig. 12: Sodium chlorate (NaClO$_3$) Raman reference spectrum (laser excitation 632.8 nm)

The most prominent band is located at a Raman shift of 935 cm^{-1} caused by symmetric stretch vibration of ClO$_3^{(-)}$. The structural similarity of sodium and potassium chlorate (KClO$_3$) is reflected in the Raman spectrum (Fig. 13) where the most intense band is situated at a relative wavenumber of 938 cm^{-1}.

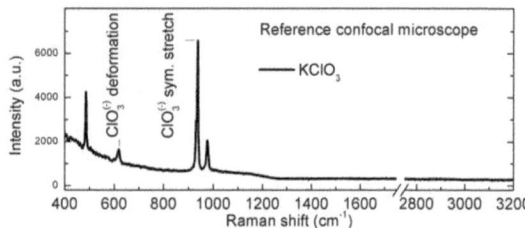

Fig. 13: Potassium chlorate (KClO$_3$) Raman reference spectrum (laser excitation 632.8 nm)

When spectroscopic systems with a reduced spectral resolution are used the band shift of 3 cm^{-1} (compared to NaClO$_3$) might not be noted. However, to differentiate between these two substances, is of only minor significance since the threat of both substances is similar.

1.6.4 Trinitrotoluene (TNT) and dinitrotoluene (DNT)

A Raman spectrum of TNT (trinitrotoluene) is shown in Fig. 14. Among the various Raman bands of TNT the symmetric C-N stretch vibration with a Raman shift of 1370 cm^{-1} is the most intense spectral feature.

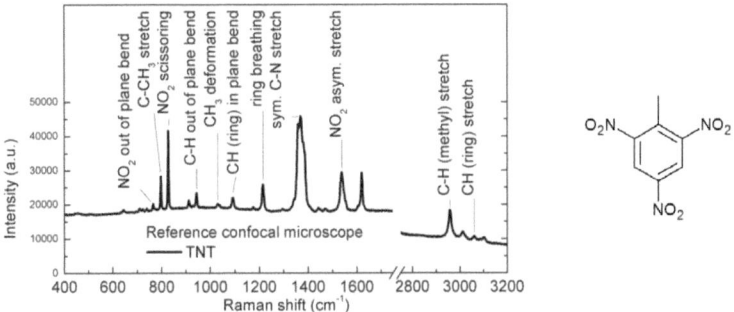

Fig. 14: Trinitrotoluene (TNT) Raman reference spectrum (laser excitation 632.8 nm) and molecular structure

The band assignment in Fig. 14 is based on a theoretical study of TNT vibrations [79]. TNT is a widely used military explosive with convenient handling properties. A degradation product of TNT is DNT (dinitrotoluene) which is also used for the production of landmines. In Fig. 15 the most prominent Raman band of DNT is located at 1365 cm^{-1}.

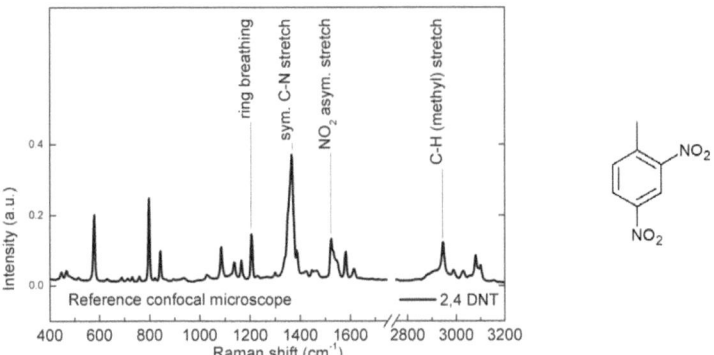

Fig. 15: 2,4-Dinitrotoluene (DNT) Raman reference spectrum (laser excitation 632.8 nm) and molecular structure

The structural similarity of DNT and TNT is reflected by various Raman bands which are at comparable spectral positions.

1.6.5 RDX and HMX

RDX (Research Department Explosive) is a military, high explosive often stabilised in plastic binders. In Fig. 16 the elevated baseline is caused by the fluorescence of the plasticiser. In addition to the N-N and the NO_2 bands between 1200-1330 cm^{-1} the ring breathing mode of the molecule causes a band at 888 cm^{-1}.

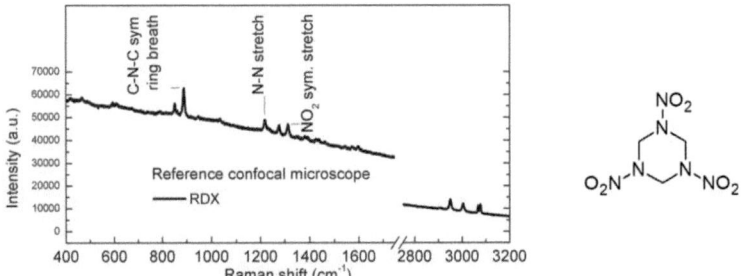

Fig. 16: Research Department Explosive (RDX) Raman reference spectrum (laser excitation 632.8 nm) and molecular structure

RDX, also known as hexogen, is the main ingredient of C4, a military plastic explosive with favourable handling properties. The use of the structurally similar HMX (also called octogen) is usually limited to special applications as it is more complicated to manufacture. The Raman spectrum in Fig. 17 contains numerous bands.

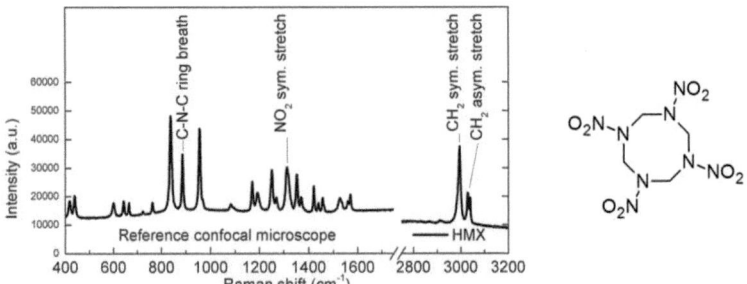

Fig. 17: HMX (octogen) Raman reference spectrum (laser excitation 632.8 nm) and molecular structure

Among various alternative names HMX stands for "her majesty's explosive".

1.6.6 Pentaerythritol tetranitrate (PETN)

PETN is one of the most powerful high explosives often used as plastic explosive. In combination with RDX it is known as Semtex. Fig. 18 illustrates the Raman bands of PETN including their assignments [80].

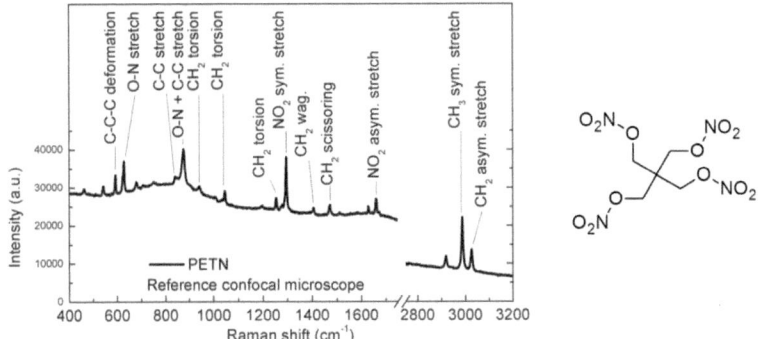

Fig. 18: Pentaerythritol tetranitrate (PETN) Raman reference spectrum (laser excitation 632.8 nm) and molecular structure

1.6.7 Hydrogen peroxide (H_2O_2) and Triacetone triperoxide (TATP)

With hydrogen peroxide improvised explosives can be synthesised, therefore the detection of this precursor is of high relevance. In Fig. 19 a Raman spectrum of a 30% aqueous solution of H_2O_2 is depicted with a intense band at a Raman shift of 877 cm^{-1}.

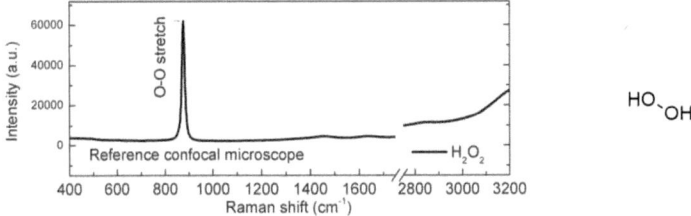

Fig. 19: Hydrogen peroxide (H_2O_2) Raman reference spectrum (laser excitation 632.8 nm) and molecular structure

When mixed with acetone H_2O_2 reacts to triacetone triperoxide (TATP), an explosive substance which is sensitive to heat, friction as well as shock. For this reason it is not used for mining but due to the ease of its synthesis it is often employed in terrorist attacks (i.e. London bombings 07/07/2005).

In Fig. 20 the spectral features of TATP are illustrated together with band assignments [81].

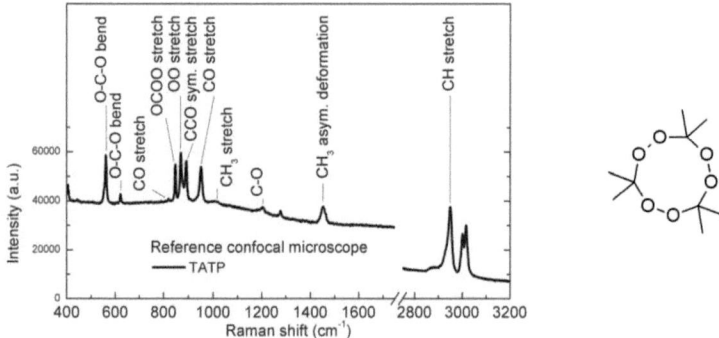

Fig. 20: Triacetone triperoxide (TATP) Raman reference spectrum (laser excitation 632.8 nm) and molecular structure

In addition to the peroxide bands between 845 and 900 cm^{-1} several bands of the organic parts of the molecule are present.

1.6.8 Diisopropyl methylphosphonate (DIMP)

Diisopropyl methylphosphonate is a yellow liquid used as a Chemical War Agent Simulant (CWAS). The Raman spectrum is given in Fig. 21.

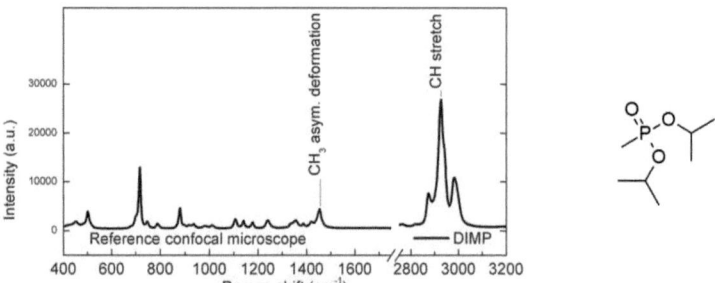

Fig. 21: Diisopropyl methylphosphonate (DIMP) Raman reference spectrum (laser excitation 632.8 nm) and molecular structure

Among the recorded bands an intensive band is located at a Raman shift of 717 cm^{-1} and at 2927 cm^{-1}.

INTRODUCTION

1.6.9 Characteristic properties of investigated explosives

The Trauzl lead block test determines the ability of a substance to widen a hole in a defined lead cylinder. Tab. 5 summarises this value together with the velocity of detonation for the analysed explosives. Furthermore, shock and friction sensitivities are listed, which are especially important parameters when handling explosives.

Tab. 5: Characteristic properties of analysed explosives [82]; velocity of detonation depends on packing density

		Trauzl lead block test ($cm^3/10\ g$)	Velocity of detonation (m/s)	Shock sensitivity (Nm)	Friction sensitivity (N)
ANFO	ammonium nitrate fuel oil	320	-	39	-
NH_4NO_3	ammonium nitrate	180	2500	49	> 353
EGDN	ethylene glycol dinitrate	620	7300	0.2	> 353
TNT	trinitrotoluene	300	6900	15	> 353
DNT	dinitrotoluene	240	-	> 50	> 353
RDX	research department explo.	480	8750	7.4	118
HMX	octogen	480	9100	7.4	118
PETN	pentaerythritol tetranitrate	523	8400	3	59
TATP	triacetone triperoxide	250	-	0.3	0.10

From Tab. 5 can be seen that the addition of fuel oil almost doubles the expansion volume of ammonium nitrate in the Trauzl test. Despite the comparatively low value of TATP in the lead block test, this substance is extremely dangerous due to its sensitivity to shock as well as friction. Another explosive sensitive to shock is EGDN. That is the reason why this substance is incorporated in a stabilising matrix for commercial use. Due to the generally good resistance of the analysed materials the applied laser energy densities used for the stand-off Raman experiments are sufficiently low to avoid detonation.

2 Setup and characterisation of stand-off system

The general principle of the stand-off Raman system will be explained first, followed by a more detailed discussion of the individual parts of the setup.

2.1 Setup overview

To obtain stand-off Raman spectra three main steps are necessary: sample excitation, light collection and spectral analysis of the Raman signal. These steps are reflected in the instrumental setup shown in Fig. 22.

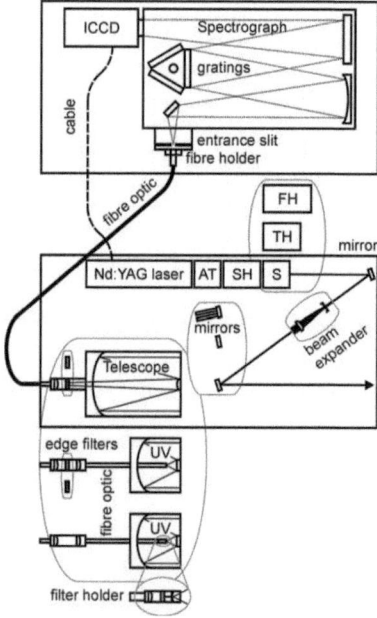

Fig. 22: Stand-off Raman setup; grey bubbles show exchangeable options for individual setups; AT attenuator, SH second harmonic, S separator, TH third harmonic, FH fourth harmonic, ICCD intensified charge coupled device

For the excitation of the samples a pulsed laser system (EKSPLA, Lithuania) providing four different wavelengths is used. The Nd:YAG laser produces laser pulses with a wavelength of 1064 nm. An attenuator (AT) allows continuous adjustment of the output power. To convert the infrared radiation (1064 nm) into green light (532 nm) a second harmonic generator (SH) is used. As the light leaving this generator consists of 1064 and 532 nm, a separator (S) removes the 1064 nm radiation. To produce even shorter wavelengths, the separator (S) is replaced by a third harmonic generator (TH) for 355 nm or a fourth harmonic generator (FH) for 266 nm. Then the laser beam is directed towards the sample using coated mirrors. Optional, a beam expander can be placed between the mirrors, consisting of two lenses. This expander widens the beam at close distance but allows to control the beam divergence at long distances. As indicated in Fig. 22, the second laser mirror can be either placed directly in front of the telescope or next to it. The advantage of the coaxial alignment is that the laser beam coincides with the axis of the telescope. As a consequence no readjustment of the laser is necessary when investigating samples at different stand-off distances. On the other hand, when placing the second mirror next to the telescope the obstruction of the telescope is minimised. This is especially advantageous when using a motorised mirror mount, which is geometrically larger than the conventional mirror holder.

To collect as much signal as possible, a telescope focuses the light returning from the sample and an edge filter removes the elastically scattered laser light, which does not contain molecular information of the sample. Different telescopes are installed to maximise the signal collection efficiency depending on the excitation laser wavelength. The collected Raman signal is focused on a fibre optic cable which guides the light to a spectrograph. This spectrograph consists of a grating which disperses the signal into its components and the intensified CCD (ICCD) camera records the spectrum. This camera is synchronised with the pulsed laser so that the measurement window coincides with the maximum Raman signal, minimising the signal contributions from sample fluorescence and daylight.

In Fig. 23 a photograph of the stand-off Raman system is depicted.

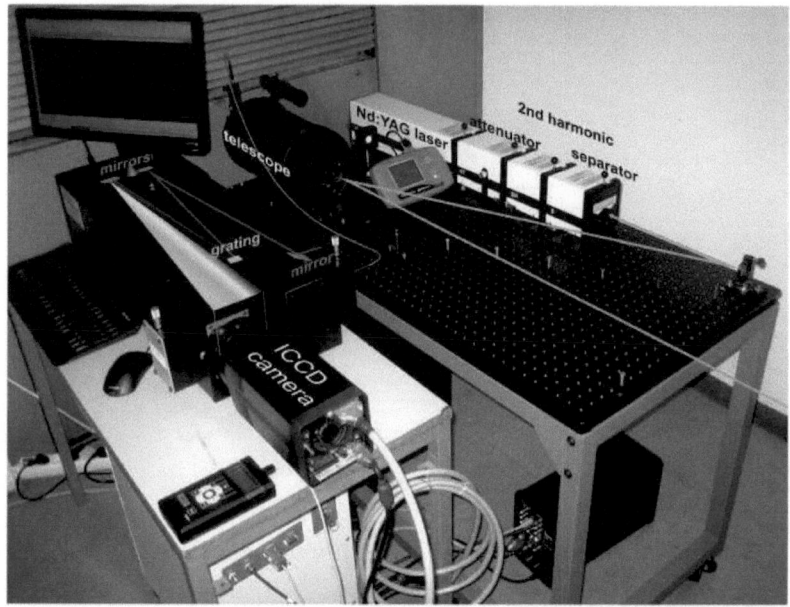

Fig. 23: Stand-off Raman system; the pulsed laser beam is directed on the sample (not in the picture). Then the telescope focuses the returning light on a fibre optic cable which directs the Raman signal into a spectrograph where the signal is separated via a grating and subsequently detected by the synchronised ICCD camera.

2.2 Setup details

2.2.1 Laser

The q-switched Nd:YAG laser (NL301HT, EKSPLA, Lithuania) provides laser pulses of the following parameters: wavelength 1064 nm, pulse length 4.4 ns, maximum average pulse energy 500 mJ, repetition rate 10 Hz, beam diameter 6 mm with a beam divergence of <0.5 mrad.

Since the well established laser technology is not common knowledge in analytical chemistry a brief overview is given here for completeness.

Fig. 24 schematically illustrates the energy levels in a Nd:YAG laser.

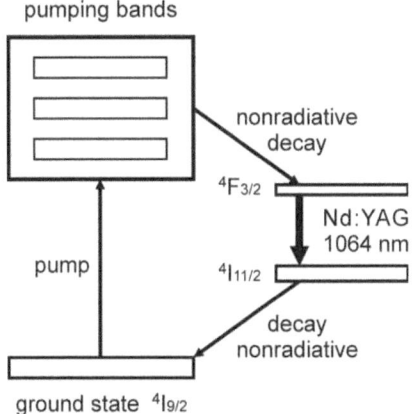

Fig. 24: Nd:YAG energy levels; picture adapted from [83]

To achieve the necessary population inversion in a four-level laser (i.e. Nd:YAG) a lower energy level ($^4I_{11/2}$) with a rapid decay is required. The higher energy state ($^4F_{3/2}$) is filled with electrons via the pumping bands which are filled themselves from the ground state ($^4I_{9/2}$). in case of the Nd:YAG laser the relaxation of electrons from $^4F_{3/2}$ to $^4I_{11/2}$ leads to the emission of 1064 nm light.

Fig. 25 shows the optical layout of the commercially Nd:YAG laser from Ekspla. Fig. 25-Fig. 30 are redrawn from those provided by Ekspla, Lithuania. The laser cavity consists of a concave rear mirror with a reflectivity of 99% and a convex output coupler. In between a Nd:YAG rod with a length of 85 mm and a diameter of 6 mm is optically pumped by a xenon flash lamp. To obtain the high energy pulses a Pockel cell is placed in the laser cavity for q-switching.

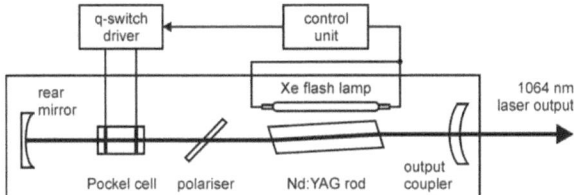

Fig. 25: Nd:YAG laser; pumped by Xe flash lamp; q-switched via Pockel cell

In order to control the output power of the Nd:YAG laser an attenuator (AT) is used. As shown in Fig. 26 the attenuator consists of two half wave plates and two polarisers. Half wave plate 2 can be rotated from the outside of the module via a micrometer screw. The rotation of this plate allows to adjust the laser output power continuously between zero and maximum energy.

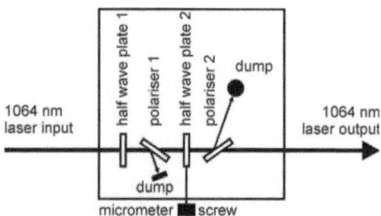

Fig. 26: Attenuator (AT); rotation of half wave plate 2 allows reduction of laser output energy

To convert the infrared radiation (1064 nm) to radiation of shorter wavelengths, nonlinear harmonic crystals are used. These crystals are mounted in adjustable holders to optimise the crystal angle for maximum output. To obtain visible, green light (532 nm) a second harmonic generator (SH) is used. The doubling of the 1064 nm input laser radiation in this module (Fig. 27) takes place in a deuterated potassium dihyrogenphosphate (DKDP) second harmonic crystal (12x12x20 mm). For the wavelength conversion, it is necessary to change the polarisation of parts of the 1064 nm input beam, which is done by a half wave plate. The output beam contains 532 as well as 1064 nm.

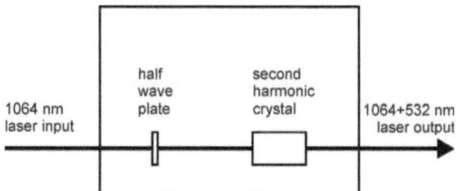

Fig. 27: Second harmonic generator (SH); frequency doubling via deuterated potassium dihydrogenphosphate crystal

To separate the two wavelengths (1064 and 532 nm) leaving the second harmonic generator a separator (S) is employed. This separator (Fig. 28) consists of a dichroic mirror which is reflective for 532 nm but transmits 1064 nm.

The transmitted infrared radiation is stopped in a beam dump, whereas the visible radiation leaves the separator via mirror 2.

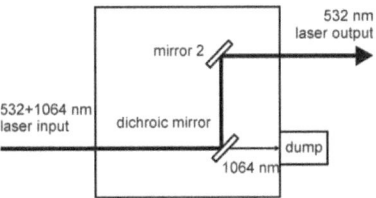

Fig. 28: Separator (S); selective reflectivity of dichroic mirror allows the separation of 1064 and 532 nm radiation

If an ultraviolet laser excitation is needed the separator (S) is replaced by the third harmonic generator (TH) shown in Fig. 29. The crystal in this module is made of DKDP (deuterated potassium dihydrogen phosphate, 12x12x20 mm) and converts the two input wavelengths (532 and 1064 nm) into radiation of 355 nm. As the two input wavelengths are already perpendicularly polarised no half wave plate prior to the crystal is needed. After the crystal a dichroic mirror separates the longer wavelength (532 nm) from the ultraviolet radiation which leaves the module via mirror 2.

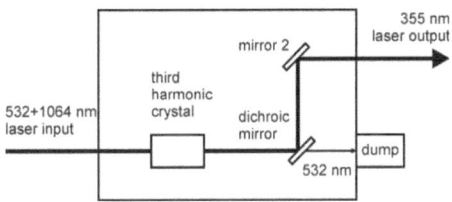

Fig. 29: Third harmonic generator (TH); DKDP crystal reduces wavelength; dichroic mirror separates radiation

To produce even deeper UV radiation a fourth harmonic generator (FH) is placed after the second harmonic generator (SH). A schematic representation of this module is given in Fig. 30. In a first step the infrared radiation (1064 nm) is reflected via dichroic mirror 1 to a beam dump. Then the beam is expanded via lens 1 and 2 before entering the fourth harmonic crystal, which is made of potassium dihydrogenphosphate (12x12x8 mm).

The separation of 532 and 266 nm is achieved by the dichroic mirror 2 and mirror 3.

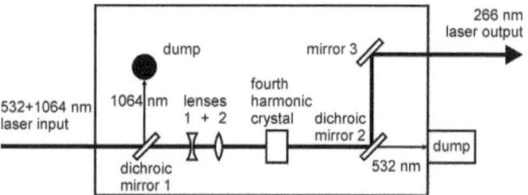

Fig. 30: Fourth harmonic generator (FH)

To characterise the laser output, the laser pulse energy, the pulse shape over time and the beam diameter were determined. The laser power was measured with a laser power and energy meter (UP19K-15S-VR, Gentec). After the stabilisation of the measurement head, an average over a time period of 20 seconds was established to obtain the pulse energy as well as the standard deviation of the laser radiation.

2.2.1.1 Temporal pulse shape

To measure the laser pulse length a photo diode (DET10A, High Speed Silicon Detector, Thorlabs, Germany) with a specified rise/fall time of 1 ns was used. 1000 laser pulses were averaged to get a reliable representation of the pulse shape. The data for 266, 355 and 532 nm is summarised in Fig. 31.

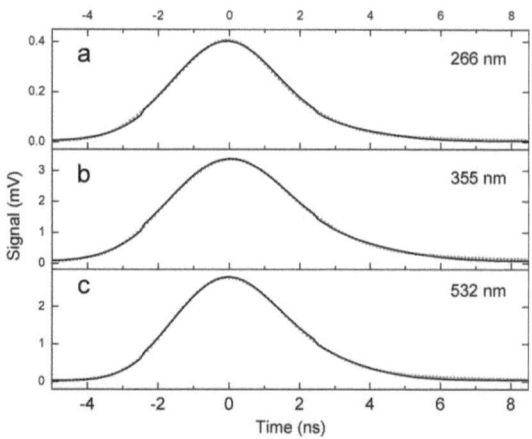

Fig. 31: Laser pulse shapes of different wavelengths; averages of 1000 pulses are shown; pulse length defined by 1/e criterion (time between the points where the signal is 36.8% of the maximal signal); pulse lengths: 266 nm (a): 4.6 ns, 355 nm (b): 5.1 ns and 532 nm (c): 4.4 ns

The data shown in Fig. 31 was used to determine the pulse shape for different laser wavelengths. Therefore, the points where the laser power was 36.8% (1/e) of the maximum power were defined as start and end of the individual pulses. The determined pulse lengths of 4.6 ns (266 nm), 5.1 ns (355 nm) and 4.4 ns (532 nm) agree well with the specified value of 4.4 ns.

2.2.1.2 Spatial beam profile

The spatial 532 nm-laser beam profile 50 cm from the laser head was determined via razor blade method. Therefore, a razor blade mounted on a linear micrometer stage was used to block the laser beam. At first, the whole beam is blocked and no light reaches the power detector, which is positioned behind the razor blade. Then, the blade was moved out of the beam in steps of 480 µm. The increase of measured pulse energy is depicted in Fig. 32a. To obtain the intensity distribution along the beam diameter the pulse energy was derived with respect to the razor position. The respective data points are shown in Fig. 32b.

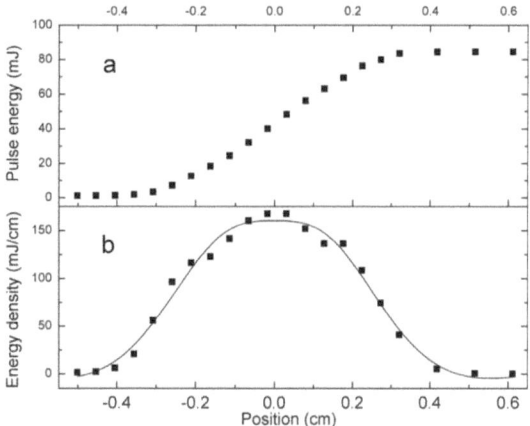

Fig. 32: Determination of 532 nm-laser beam diameter via razor blade method; a: measured pulse energy increases as the blade is moved out of the beam; b: derived power distribution along the laser beam diameter; pseudo Voigt fit with an adj.R^2 of 0.983 and a FWHM of 5.3 mm

The energy density along the beam diameter was approximated with a pseudo Voigt regression, shown as a black line in Fig. 32. The Full Width at Half Maximum (FWHM) is 5.3 mm. The alternative 1/e criterion results in a beam width of 6.07 mm.

For a more detailed determination of the spatial laser profile a power meter was used to manually align both laser cavity mirrors for maximum IR power output. Then the laser beam profile of a single laser pulse was determined using burn paper at a distance of 10 cm from the laser output. Fig. 33 illustrates the symmetric energy distribution across the beam diameter.

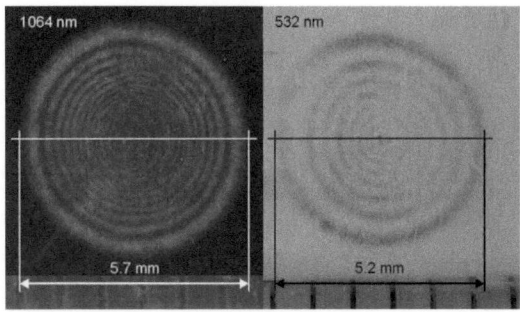

Fig. 33: Beam profiles of single laser pulses recorded 10 cm from the laser head; 1064 nm: energy 319 mJ±0.8mJ, diameter 5.7 mm; 532 nm: energy 113 mJ±0.5mJ, diameter 5.2 mm

After the optimisation of the 1064 nm-beam, the second harmonic crystal was placed in front of the Nd:YAG laser module. The angle of the second harmonic crystal was optimised for maximum power output and the beam profile was recorded on a thermo sensitive paper, leading to a symmetric laser profile (Fig. 33, right) which shows good agreement with the profile determined via the razor blade method.

In order to expand the beam size at remote distances a beam expander can be used. The expander shown in Fig. 34 consists of a plano-concave lens (a) with a focal length (f) of 50 mm. When the plano-convex lens (b) with a focal length of 150 mm is placed at a position where the focal points (F) of both lenses coincide, a beam with a diameter (d) of 6 mm is expanded to a diameter of 18 mm.

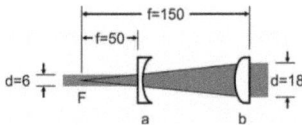

Fig. 34: Galilean beam expander to control laser beam divergence; the beam diameter (d) is expanded according to the ratio of the focal lengths (f) of both lenses (a, b), when focal points (F) coincide

The built beam expanders exhibits a Galilean configuration, where the beam is expanded via a plano-concave lens before re-collimation with a plano-convex lens. Alternatively, an expansion can be achieved via two convex lenses as well. However, with high pulse energies it is important to avoid this Keplerian configuration, as the laser goes through a focus where the energy density can get high enough to ignite an air plasma. In Fig. 35a the average beam profile of 600 green laser pulses is shown at a distance of 100 m. For the beam profile of a single laser pulse at a distance of 20 m, (depicted in Fig. 35b), no beam manipulation was undertaken.

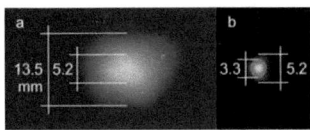

Fig. 35: Beam profile at 532 nm; a: average of 600 laser pulses at 100 m, pulse energy at the laser head 217 mJ; b: 50 mJ-single pulse at 20 m

Fig. 35a illustrates that the beam expander allows one to control the divergence of the laser beam at further distances. But a pair of lenses can not only be used to produce expanded beams which are collimated. By changing the distance between the two lenses the laser beam can be focused at a particular distance. For a distance of 40 m the resulting burn spots on a black photo paper are shown in Fig. 36. Therefore, varying numbers of 532 nm laser pulses with a pulse energy of 13.4 mJ where accumulated.

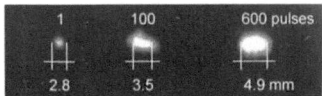

Fig. 36: Focused beam profiles at 40 m with varying pulse numbers per spot; increasing pulse accumulation shows the point stability which is limited by atmospheric fluctuations

In Fig. 36 the spot size of the focused laser beam can be seen on the left, where only one single laser shot was applied to the target. With increasing numbers of laser pulses the laser point stability leads to a wider discolouration of the photo paper.

To record stand-off Raman spectra it is of great importance to control the beam size as well as the position of the laser spot on the sample. These parameters are of particular importance for stand-off Spatial Offset Raman spectroscopy (SORS), a technique which is detailed in a later chapter.

2.2.2 Mirror mounts

Depending on the specific necessities for an experiment, alternative mirror positions are advantageous. If several samples at different stand-off distances should be measured the coaxial arrangement minimises setup re-adjustment between the samples. For this alignment the last mirror is placed in front of the telescope to bring the laser beam in the telescope axis. This means, only the focus of the telescope and the delay time between laser pulse and ICCD camera gate have to be re-optimised when measuring a substance at a varied distance. However, the disadvantage of the coaxial configuration is the obstruction of the telescope by the mirror. If the secondary mirror of the telescope is bigger than the laser mirror, no additional obstruction of the telescope is caused. However, for precise laser positioning a motorised mirror mount is used (stepper motorised ultra-stable kinematic Ø1" Mirror Mount, Thorlabs) which is larger in size than a manual kinematic mirror mount. Therefore, the mirror was placed next to the telescope. With this oblique geometry no obstruction of the collection optic occurs. The motorised mirror mount is moved by two linear stepper motors which are positioned at two corners of the mount, as illustrated in Fig. 37.

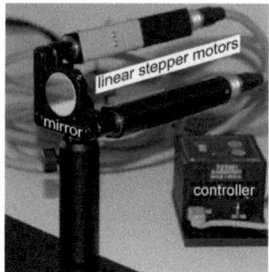

Fig. 37: Motorised mirror mount with two linear stepper motors for two dimensional laser positioning

The mirror mount is constructed in such a way that the movement of only one motor changes the mirror angle neither only horizontally nor only vertically. A combined motion of both motors is needed to achieve pure vertical or horizontal movement of the mirror. Therefore, a bilinear calibration algorithm was developed in cooperation with the Institute of Geodesy and Geophysics of the faculty of Mathematics and Geoinformation at the Vienna University of Technology and implemented in a graphical user friendly interface in LabView. To determine the relation between motor and mirror movement the motors automatically move to eight specified positions. The horizontal as well as the vertical distances between the eight corresponding laser spots are measured at the sample distance. With these coordinates (x and y) and the known motor positions (S_x and S_y) the eight parameters (a_{0-3} and b_{0-3}) in the two following equations (Equation 1) are calculated.

$$S_x = a_0 + a_1 x + a_2 y + a_3 xy$$
$$S_y = b_0 + b_1 x + b_2 y + b_3 xy$$

Equation 1: Correlation of motor positions (S_x and S_y) and laser coordinates on the sample (x and y); a_{0-3} and b_{0-3} are parameters calculated via bilinear calibration

With this established bilinear calibration, the necessary motor positions (S_x, S_y) are calculated to aim the laser at certain spots (x, y) on the sample.

To show the precision of the motorised mirror mount the focused laser spot was directed to different positions on a photo paper at a distance of 40 m Fig. 38.

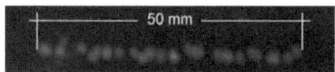

Fig. 38: Line scan of focused laser spots on photo paper at 40 m distance; achieved with motorised mirror mount; average distance between consecutive burn spots is 1.75 mm

To achieve the distance of 1.75 mm between the individual laser spots on the target, the stepper motor was moved 0.8 µm.

2.2.3 Telescopes

The probability of a photon to undergo inelastic scattering is on the order of only $1:10^7$ [84]. Furthermore, the Raman intensity decreases with increasing sampling distance since the Raman photons are scattered in all directions. These are the reasons why telescopes are used to collect as many Raman photons emitted from the sample as possible. Depending on the laser excitation wavelength, different telescopes are used to maximise the light collection efficiency. For the collection of light in the visible or near UV range a Schmidt-Cassegrain telescope (Celestron C6) with a clear aperture of 152.4 mm (6 inches) was used. With a focal length of 1500 mm the half angle of the focused light is 2.9°, small enough for the edge filter to remove the elastically scattered laser light (edge filters are discussed in the next section). The mirrors are made of Al, SiO_2 and TiO_2, however due to the Schmidt plate, which covers the entire telescope aperture, this telescope is not suitable for the collection of Raman spectra when an excitation laser of 266 nm is employed. Therefore, a Schmidt telescope (McPherson) with a focal length of 125 mm (5 inches) and an aperture of 125 mm is used for the collection of UV radiation. In this case the mirrors are made of Al and MgF_2 and due to the unprotected mirror coating the UV reflectivity is maximised. Fig. 39 shows pictures of both telescopes.

Fig. 39: Telescopes; a: 6" telescope (Celestron) for visible and near-UV radiation; b: 5" UV telescope (McPherson)

2.2.3.1 Mechanical adaption of UV telescope

For measurements in the UV it is necessary to use collection optics which are suitable for this spectral region. Therefore, a UV telescope was purchased from McPherson. The mirrors were made of aluminium and magnesium fluoride to maximise the reflectivity in the UV. In Fig. 40 the UV Schmidt telescope with a focal length of 125 mm and an aperture of 125 mm is depicted.

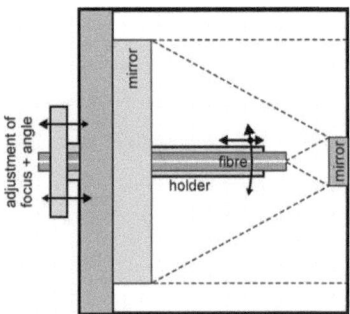

Fig. 40: McPherson telescope; 100 µm-fibre optic cable in holder for fibre adjustment; angle and axial position can not be adjusted independently

From Fig. 40 it can be seen that the collected light is focused within the telescope. Therefore, a fibre optic cable with a diameter of 100 µm was placed in the telescope.

The fibre was mounted in an adjustable holder which allowed two movements. Firstly, to adjust the system for signal collection from a particular distance, the holder with the fibre could be moved further in or out of the telescope. This was achieved with four screw pairs at the back of the telescope. Secondly, the fibre could be tilted to collect light which did not originate exactly from the telescope axis. For this tilt adjustment, however, the same four screw pairs had to be used. This means, in reality, it was close to impossible to adjust the fibre for a different stand-off distance without unintentionally changing the tilt angle.

Together with the Institute for Production Engineering and Laser Technology at the Vienna University of Technology, the fibre holder was improved to allow independent movement along the axis and adjustment of the fibre tilt angle. This was achieved by an additional metal plate at the back of the telescope for angle adjustment (Fig. 41). For the improved holder the screw pairs were reduced from four to three, to guarantee statically determined adjustment of the tilt angle.

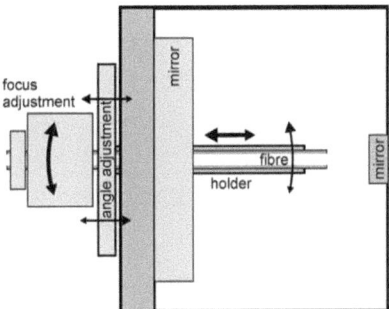

Fig. 41: UV telescope with modified fibre holder for independent adjustment of the fibre tilt angle and along the telescope axis

To focus at particular object distances the fibre is moved along its axis by turning a big handle at the back of the telescope (Fig. 41). With the reconstructed holder this alignment does not change the adjusted fibre angle relative to the telescope.

2.2.3.2 Signal from fibre and laser

To evaluate the efficiency of the modified telescope, stand-off Raman spectra were recorded at a distance of 4.2 m accumulating 300 laser pulses with a pulse energy of 5 mJ and a wavelength of 266 nm.

As expected, the nylon spectrum (Fig. 42a) showed bands between 1250-1750 cm^{-1} and 2650-3150 cm^{-1} however additional bands at 614, 810 and 1066 cm^{-1} were present. Via confocal Raman microscope (Fig. 42b) these additional bands were identified as quartz bands caused by the fibre optic cable (UV/VIS fibre Ø1000 µm, 1 m long, Avantes, Netherlands). As a conclusion, the collected laser light was sufficiently intense to excite Raman signals in the fibre which then transmitted through the filter which was positioned after the quartz fibre. To suppress the generation of these interfering fibre bands the filter was placed in front of the fibre optic within the UV telescope (McPherson) Fig. 42c+d.

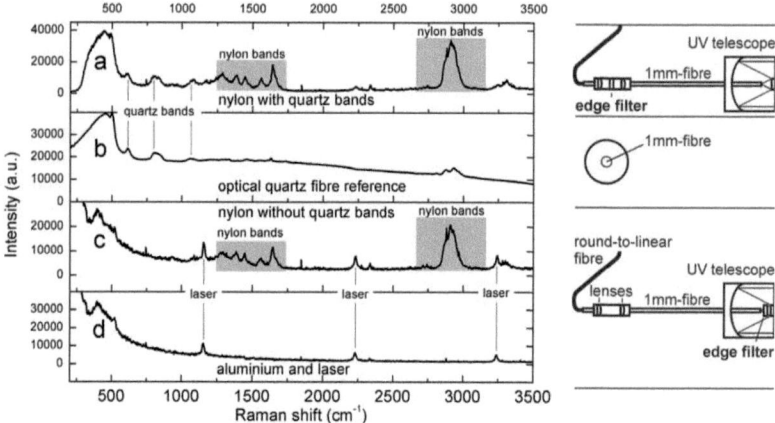

Fig. 42: Raman interference bands of fibre optic cable; a: stand-off nylon spectrum; edge filter placed after optical fibre; fibre quartz bands between 600 and 1100 cm^{-1}; b: fibre optic comparison spectrum via confocal Raman microscope; c: stand-off nylon spectrum; edge filter in front of optical fibre; no fibre bands but laser light passes the filter; d: stand-off aluminium spectrum shows laser artefacts as well

Due to the new filter position, in Fig. 42 the fibre quartz bands are neither present in the nylon (c) nor the aluminium spectrum (d), however laser artefacts are visible. The reason for this behaviour is the low ratio of focal distance and aperture of the telescope. With a ratio of only 1 the light is focused at a half angle of 26.6°, not sufficiently collimated for the filter with an acceptance angle of ±2°. To avoid Raman excitation of the fibre optic as well as to guarantee effective removal of the laser light, a filter adapter was developed as described in the following section.

2.2.3.3 Filter adapter for 266 nm

Raman quartz bands from the optical fibre were recorded in the spectra when using a laser excitation wavelength of 266 nm. Therefore, it is necessary to place a Raman edge filter in front of the fibre, rather than after the fibre in the optical collection assembly. Since the fibre optic is located in the UV telescope, the limited space required the design and the construction of a miniaturised filter holder, as illustrated in Fig. 43.

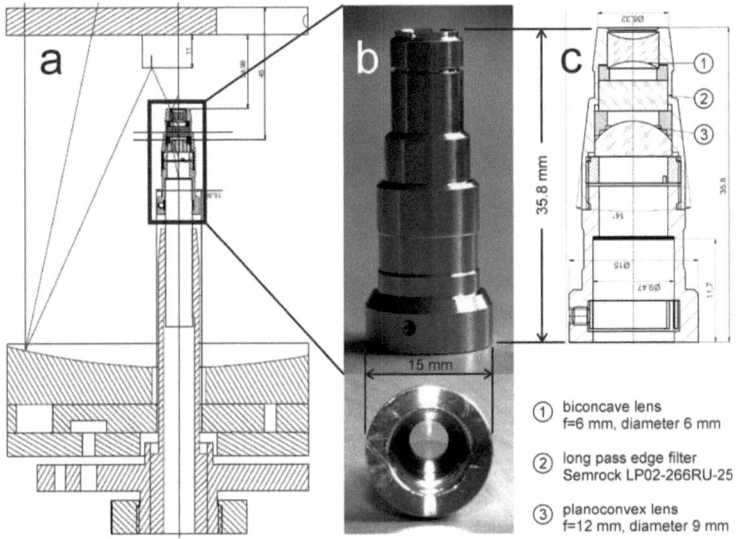

Fig. 43: Raman edge filter optic adapter for UV telescope at 266 nm laser excitation wavelength; made by Schäfter und Kirchhoff, Hamburg, Germany; a: drawing of UV telescope with filter holder (rectangle); b: photograph of filter holder; c: drawing of filter holder

With a total length of 35.8 mm this adapter can be mounted on the optical fibre within the UV telescope. It contains a long pass edge filter with a cut off wavelength of 266 nm (Semrock, LP02-266RU-25). To work efficiently this filter requires collimated light with an acceptance angle of ±2°. Therefore, a biconcave lens (focal length 6 mm, diameter 6 mm) collimates the focused light from the telescope (focal length 125 mm, diameter 125 mm). After the collimated light has passed the edge filter a plano-convex lens (focal length 12 mm, diameter 9 mm) refocuses the beam on the optical fibre, which guides the light into the spectrograph.

2.2.4 Edge filters

Since only a minute portion of the incident laser photons undergo inelastic scattering it is important to remove the elastic scattered part of the collected radiation. This is achieved by using high pass edge filters (Semrock) for each of the three used laser excitation wavelength (266, 355 and 532 nm). In Fig. 44 the absorption spectra of the edge filters are shown, recorded on a UV/VIS spectrometer (Hewlett Packard 8452A).

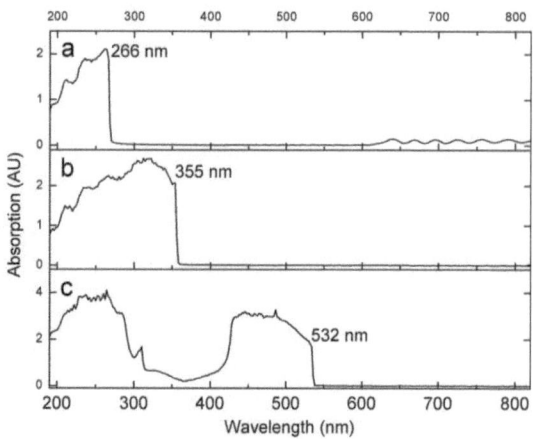

Fig. 44: Absorption spectra of Raman high pass edge filters for a: 266, b: 355 and c: 532 nm

In Fig. 44 the effect of the used edge filters is visualised. Whereas they transmit light above the individual cut off wavelength, they block the light at shorter wavelengths. In this way the excitation laser wavelength is blocked and the red shifted Stokes signals can penetrate the filter.

2.2.5 Fibres

The collected signal is focused on a fibre optic cable which guides the light to the spectrograph. Compared to direct coupling of telescope and spectrograph the throughput of fibre coupling is about ten times lower [85], however, the flexibility in system design is much improved [8]. Therefore, a fibre optical bundle of 19 200 µm-fibres was used (Avantes, Netherlands), with a length of 1 m and a numerical aperture of 0.22. The spatial filtering at the entrance slit of traditional spectrometers leads to a trade of between spectral resolution and throughput [86]. To optimise both parameters the fibre optical bundle was designed in a round-to-linear configuration. This means that the 19 single fibres are aligned in round geometry at the entrance, whereas they exhibit a linear alignment at the fibre exit, as illustrated in Fig. 45.

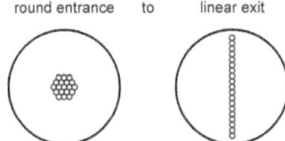

Fig. 45: Fibre optical cable with 19 individual 200 µm-fibres; round entrance and linear exit configuration to optimise throughput and spectral resolution

To increase the light entering the round fibre end, a plano-convex lens is placed in front of the fibre, to match the numerical aperture of the fibre. The linear fibre exit is mounted in a bespoke fibre holder which allows adjustment in two dimensions, for ideal fibre placement in front of the spectrometer entrance slit. The described configuration is used with the 6'' Schmidt-Cassegrain telescope (Fig. 46 a). However, when the 5'' Schmidt UV telescope is used, a single 1 mm-fibre (Avantes, Netherlands) collects the focused light in the UV telescope (Fig. 46 b). At the exit of this fibre (length 1 m, numerical aperture 0.22) a plano-convex lens (diameter 6 mm, focal length 8.7 mm) collimates the light before it passes the edge filter. After the filter an identical lens focuses the light into the round-to-linear fibre, which guides the signal to the spectrograph.

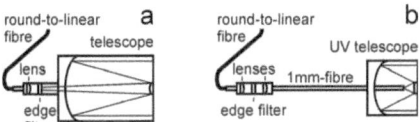

Fig. 46: Fibre arrangements to guide focused signal from the telescope to the spectrograph; a: 6'' Schmidt-Cassegrain telescope; b: 5'' Schmidt telescope for the UV range

2.2.6 Spectrograph

An imaging Czerny-Turner spectrograph (Acton SP-2750) with a focal length of 750 mm is used to separate the signal, using a triple grating turret, equipped with three gratings; two ruled gratings with a blaze wavelength of 500 nm (300 and 1800 grooves/mm) and a holographic grating optimised for UV light (2400 grooves/mm). With a grating size of 68x68 mm, the ratio of focal length and aperture is 9.8.

2.2.7 ICCD camera

At the back of the spectrograph an ICCD (Intensified Charge Coupled Device) camera (Princeton Instruments) is attached with an adjustable exposure time down to 500 ps. The 1024x256 CCD chip (pixel size 26 µm) is thermo-electrically cooled to -20°C. The spectrograph disperses the signal along the horizontal chip axis and the chip is binned column wise to obtain the spectrum. The Princeton Instruments WinSpec software package from Roper Scientific provides two different data acquisition modes. Usually the detected light from multiple laser events is accumulated on the CCD chip before the data is read out. To avoid detector saturation when measuring intense Raman scatterers or fluorescent samples, the data accumulation can be achieved by mathematically summing multiple camera read outs. All data in this work is shown as observed, neither baseline adjustment nor cosmic ray correction was applied. In Fig. 47 the radiation collected from a fluorescent light bulb was dispersed along the horizontal axis of the CCD chip via a 300 grooves/mm-grating. The circular contour in Fig. 47 is a result of the camera intensifier with a diameter of 18 mm.

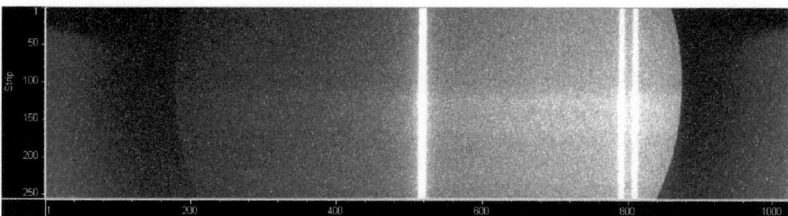

Fig. 47: Signal distribution on the CCD chip; the signal of a fluorescent light bulb is depicted; the circular contour indicates the 18 mm diameter of the camera intensifier

The three bands of higher signal intensity visible in Fig. 47 correspond to emission lines of the fluorescent light bulb.

Summing up the intensities column wise leads to spectral information in a more usual representation, as can be seen in Fig. 48 where the intensity is plotted along the vertical axis, whereas the horizontal axis states the CCD column number.

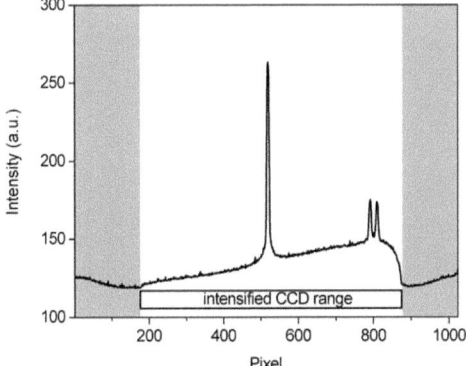

Fig. 48: Spectrum of fluorescent light bulb along one CCD pixel row; only the middle area of the chip is intensified

In Fig. 48 the spectral base line is particularly elevated in the middle of the CCD chip (pixels 180-860) where the intensifier functions. The reason for that is the size of the intensifier (18 mm) which is smaller than the CCD chip. Furthermore, the pixel number stated on the horizontal axis is not helpful when it comes to substance identification. Therefore, the spectrograph needs calibrating, with a light source of known emission. Preferably an irradiation source with several narrow bands is used. The three bands visible in Fig. 48 can be identified in the reference spectrum shown in Fig. 49 at 546.7, 577.5 and 579.7 nm.

As the band positions are well known, the CCD pixel position can be replaced with the wavelength in nm, allowing chemical identification.

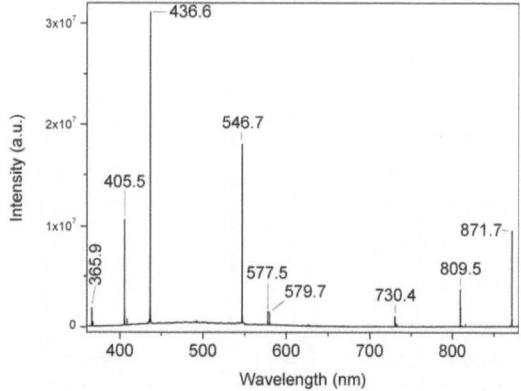

Fig. 49: Spectrum of fluorescent light bulb for calibration; numbers in diagram state band wavelengths in nm

A closer look on the spectral data of the light bulb, Fig. 50 reveals a repeated band pattern at higher wavelengths. These lines are the second diffraction order of the light emission, generated by the diffraction grating in the spectrograph. One can derive the position of a second order band by multiplying the wavelength of the first order signal.

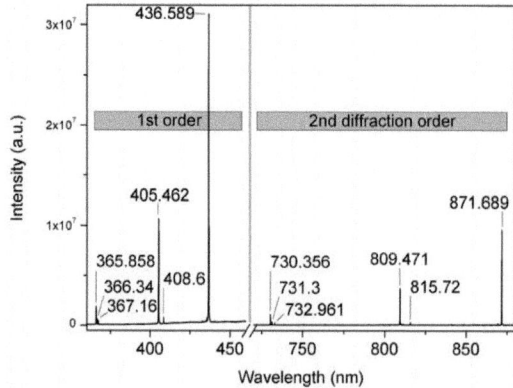

Fig. 50: 1st and 2nd order spectral lines of fluorescent light bulb; numbers indicate band wavelengths in nm

Due to this effect an overlap of first and second order diffractions can occur when recording spectral data over an extended wavelength range. In Raman spectroscopy the measured wavelength range is small enough and the second order spectrum is not detected as shown in Fig. 51.

On the horizontal axis of Fig. 51 the calculated Raman shift is given in wavenumbers (cm^{-1}) relative to the excitation wavenumber of the laser. On the vertical axis the detected signal wavelength is stated in nm.

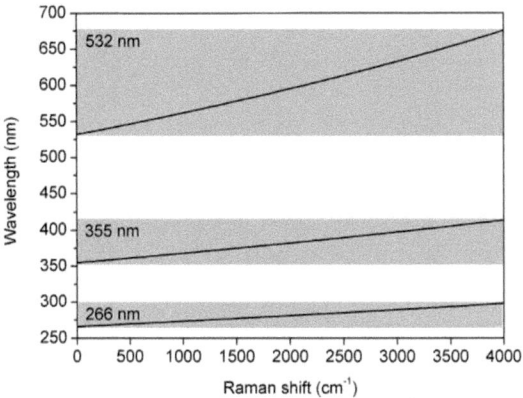

Fig. 51: Detected wavelength in nm vs. calculated Raman shift in cm^{-1} for three laser excitation wavelengths stated in the diagram

The grey bars indicate the measured spectral range in nm on the vertical axis when investigating Raman signals between 0 and 4000 cm^{-1}. Excitation with a green (532 nm) laser leads to signals between 532 and 675.8 nm. With an excitation in the near UV (355 nm) the same signals range between 355 and 413.8 nm. Further reduction of the excitation wavelength to 266 nm shifts the bands to 266-297.7 nm.

As stated in a previous paragraph Fig. 51 shows that higher diffraction orders will not be present in the typical Raman measurement range. Fig. 52 shows 15 m-stand-off spectra of $NaClO_3$. For each of these spectra a different laser excitation wavelength was employed, indicated below the individual spectrum, together with the laser pulse energy and the number of accumulated laser pulses.

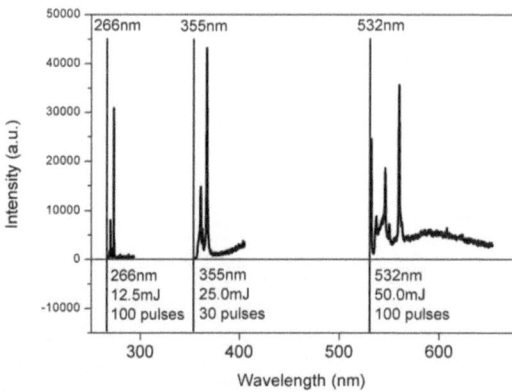

Fig. 52: Influence of excitation laser wavelength; 15 m-stand-off Raman spectra of solid NaClO$_3$; excitation wavelength, laser pulse energy and number of accumulated laser pulses are stated below the spectra

For all three laser wavelengths the same spectral region (Raman shift 0-3500 cm^{-1}) of NaClO$_3$ was recorded. When the spectral data is represented in nm the spectra of NaClO$_3$ is located at different positions, depending on the excitation laser wavelength. When reducing the laser excitation wavelength the Raman spectrum is spread across a much smaller nm-wavelength. The spectral resolution of the setup at a Raman shift of 2000 cm^{-1} is listed for different laser excitation wavelengths and gratings in Tab. 6.

Tab. 6: Setup resolution in cm^{-1} (FWHM) at different laser excitation wavelengths (532, 355 and 266 nm) with different gratings for a centre wavelength of 2000 cm^{-1}; spectrograph SP-2750 and PI MAX1024-18-GenII ICCD camera; 'out of range' refers to the mechanical rotation limitation of the grating in the spectrograph

grating grooves/mm	532 nm laser excitation (2000 cm^{-1} = 595.3 nm)	355 nm laser excitation (2000 cm^{-1} = 382.1 nm)	266 nm laser excitation (2000 cm^{-1} = 280.9 nm)
300	9.89 cm^{-1}	24.15 cm^{-1}	44.79 cm^{-1}
1800	1.31 cm^{-1}	3.65 cm^{-1}	7.05 cm^{-1}
2400	out of range	2.54 cm^{-1}	5.08 cm^{-1}

To maintain good spectral resolution when exciting with UV lasers gratings with more grooves per millimetre are required, as can be seen in Tab. 6.

2.2.7.1 Influence of ICCD gain on signal intensity

The employed ICCD camera allows the amplification of collected signals. To find the optimum camera gain 100 m-stand-off Raman spectra of a 25 mm thick polypropylene (PP) sheet were recorded with the intensifier gain varied between 0 and 255.

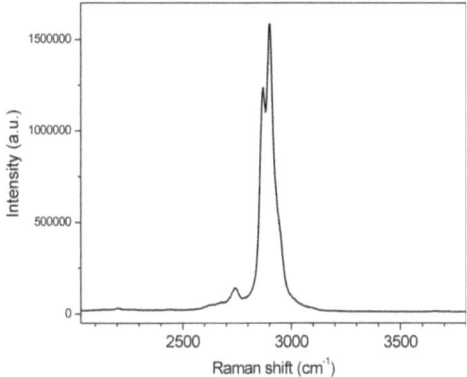

Fig. 53: 100 m-stand-off Raman spectrum of polypropylene (25 mm thick) with a ICCD gain of 128; laser wavelengths 532 nm, pulse energy 213 mJ, accumulation of 100 laser pulses

From the obtained spectra the baseline corrected signal intensity at 2852 cm^{-1} and the RMS noise (standard deviation of the spectral data points) between 3500 and 3600 cm^{-1} was determined. The results are presented in Fig. 54 together with the calculated Signal-to-Noise ratio (S/N).

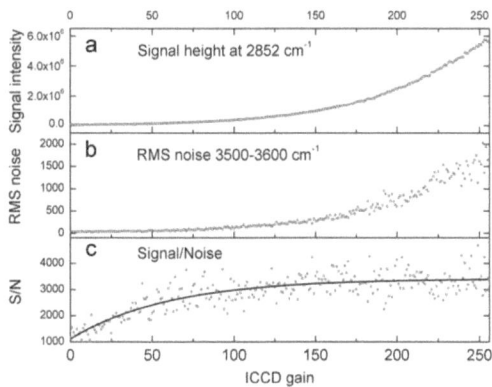

Fig. 54: Influence of ICCD camera gain exemplified on a 25 mm PP sheet at 100 m ; a: signal height at 2585 cm^{-1}; b: RMS noise between 3500 and 3600 cm^{-1}; c: signal to noise ratio

With increasing gain the signal rises significantly (Fig. 54a). However, as the same is true for the noise level (b) the calculated ratio of signal and noise is depicted in Fig. 54c. The S/N increases with the ICCD gain at low values, but levels off at a medium range. Therefore, a ICCD gain factor of 128 was used for all experiments.

2.2.7.2 Gate time influence on recorded signal

It was shown that the signal to noise ratio of a CCD camera compared to an ICCD camera is of a similar magnitude [8]. But the synchronisation of the short camera gate with the laser pulse allows the rejection of ambient light as well as fluorescence. To show the effect of increasing camera gate a nylon plate at a distance of 15 m was measured and the camera gate was varied from 500 ps to 25 ms. To obtain the signal-to-noise ratio the band area between 2750 and 3050 cm^{-1} was divided by the RMS noise between 2100 and 2400 cm^{-1}. Fig. 55 shows the recorded spectra (a) and the decrease in signal to noise ratio at longer gate times (b) illustrates the importance of a short gated detection system.

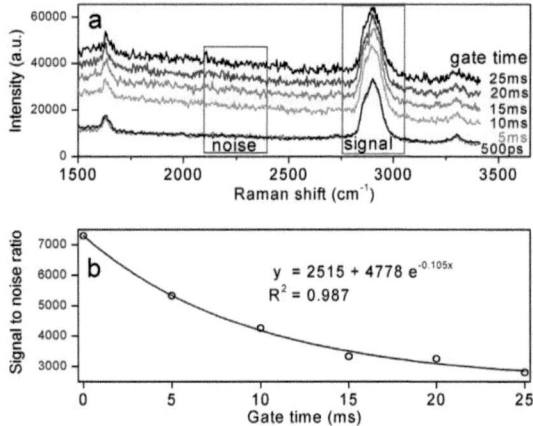

Fig. 55: Gate width impact on nylon stand-off Raman spectra; a: 15 m-spectra; 532 nm excitation; energy 50 mJ/pulse; for the 500 ps gate, 100 spectra were added to compensate for shorter collection time; for all other gate widths, 10 spectra were added. b: signal-to-noise ratio vs. gate width

2.2.7.3 Camera synchronisation with laser

In addition to a short camera gate to reject ambient light, it is of great importance to choose the optimum delay time between laser pulse and camera gate. According to the speed of light, this delay changes with the distance of a sample from the setup. Furthermore, the short gate of the camera allows to separate the instantaneous Raman signal from fluorescence, by setting an ideal delay time. Fig. 56 shows the separation of the Raman signal when analysing DIMP (diisopropyl methylphosphonate) at a distance of 5 m with a 355 nm laser [14].

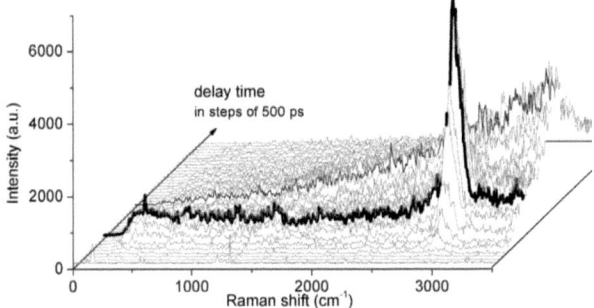

Fig. 56: Fluorescence rejection via delay time selection between laser and ICCD camera; 355 nm-stand-off Raman spectra of DIMP at 5 m; 50 mJ laser pulse energy, 10 pulses/spectrum; fluorescence maximum (thin, black) appears 5 ns after the maximum Raman signal (bold, black)

In Fig. 55 the Raman spectrum of DIMP (black, bold) appears 5 ns before the fluorescence (thin, black), allowing a temporal separation of Raman signal and unwanted fluorescence.

2.3 Influence of laser power and measurement time on signal quality

Solid ammonium nitrate (NH_4NO_3) as a pellet and diisopropyl methylphosphonate (DIMP) in a 5 mm quartz cuvette were analysed at a distance 5 m from the setup. The influence of measurement time and laser power on the recorded spectral quality was determined for a laser excitation wavelength of 532 as well as 355 nm. The laser beam diameter on the sample was 6 mm (28.3 mm^2).

Fig. 57 shows how the number of laser pulses accumulated on the CCD chip of the ICCD camera influences the recorded Raman signal of NH_4NO_3 (a) as well as the RMS noise (b).

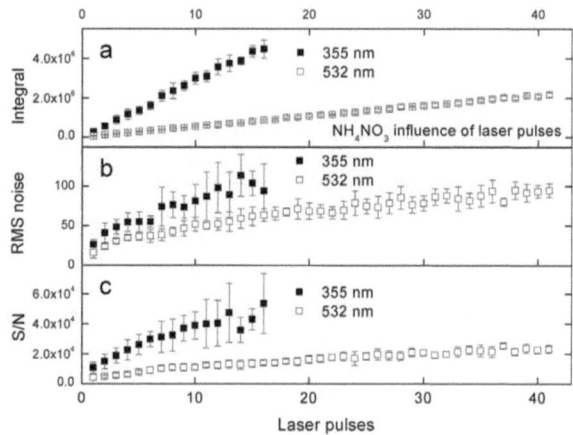

Fig. 57: Comparison of laser excitation wavelengths for the measurement of NH_4NO_3; 355 nm (black squares), 532 nm (white squares); signal integral S (a), RMS noise N (b) and S/N (c); laser pulses of 50 mJ/pulse; values are averages of nine consecutive measurements, error bars represent one standard deviation

Comparing the results of the two different excitation wavelengths (Fig. 57), the S/N (c) is significantly higher for the near UV (355 nm) rather than for the visible excitation (532 nm). Due to the increased Raman response when exciting the NH_4NO_3 sample with 355 nm, the signal accumulation was limited to a maximum of 16 laser pulses, to avoid detector saturation. Excitation with 266 nm did not further improve the detection, since the employed telescope (Celestron C6) was not suitable for the used wavelength.

Another way to modify the spectral quality is by changing the laser power. Therefore, the laser pulse energy was varied between 0 and 60 mJ. As shown in Fig. 58 the signal integral increases with the applied pulse energy for DIMP measured at 532 nm.

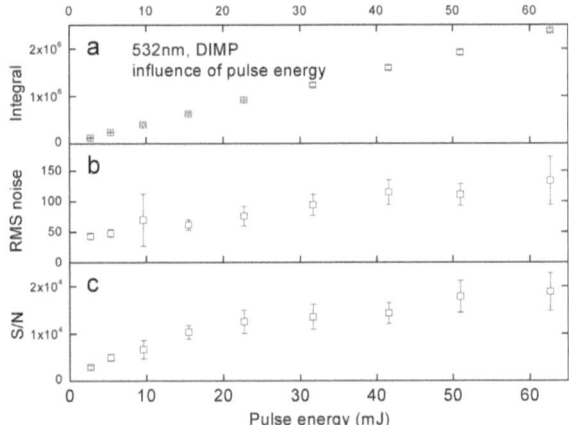

Fig. 58: Influence of laser pulses energy on signal integral S (a), RMS noise N (b) and S/N (c); DIMP at 5 m, accumulating 400 laser pulses of 532 nm; values are averages of nine consecutive measurements, error bars represent one standard deviation

In Fig. 58 the linear increase of the Raman signal (a) together with the monotonous improve of the S/N (c) show that high laser powers are of great importance for the analysis of distance objects, however bearing in mind that the sample must not be destroyed by the laser.

The signal intensity and the S/N increased with higher pulse energy and longer measurement time. This behaviour was observed for DIMP as well as for NH_4NO_3 for both excitation wavelengths.

2.3.1 Stand-off distance influence on delay and signal intensity

To investigate Raman signal intensities with increasing distance, a 25 mm thick polyethylene (PE) sheet was used as a sample. The sample was placed at different stand-off distances ranging from 10 to 100 m in steps of 10 m. At distances from 10 to 40 m the laser output pulse energy was set to 50 mJ. For distances between 50 and 100 m the pulse energy was increased to 223 mJ to compensate for atmospheric losses. For each distance the measurement system was optimised for maximum signal collection efficiency and 600 pulses were added for each sample distance. In Fig. 59 the resulting PE spectra at different stand-off distances are shown.

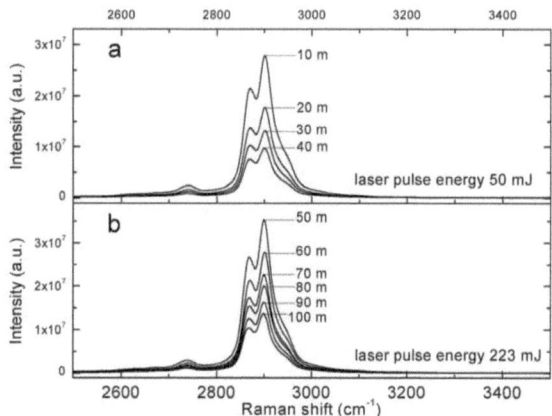

Fig. 59: Stand-off Raman spectra of a 25 mm-PE sheet at different distances, stated in the diagram; laser pulse energy: (a) 50 mJ, (b) 223 mJ

Fig. 59a shows the PE spectra measured between 10 and 40 m, whereas the panel b summarises the data recorded between 50 and 100 m. As Raman scattered light does not leave excited matter in a preferred direction, the number of photons entering the telescope, decreases with the distance from the sample. In order to establish a quantitative relationship between distance and recorded signal, the PE band was integrated between 2400 and 3400 cm^{-1} and the baseline subtracted.

Taking into account the linear increase of Raman signal with higher excitation energy, as described in a previous section, the determined integrals were divided by the laser pulse energy measured at the laser head in millijoule. The resulting values are shown in Fig. 60

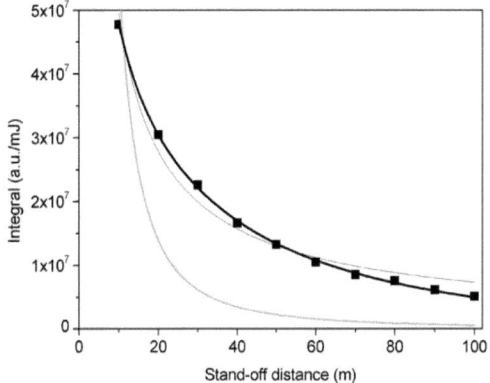

Fig. 60: Change of collected PE Raman signal with stand-off distance; 3 alternative fitting methods
black squares: baseline corrected PP integrals (2400-3400 cm^{-1}) divided by laser energy in mJ
bold black line: Integral=1.57E8/distance^0.46*exp(-0.0134*distance), adjusted R^2=0.9995
grey line: 1/r^2 'fit', uniform scattering in all directions; Integral=5.5E9/distance^2
thin black line: flexible power 1/r^p; Integral=3.3E8/distance^0.83, adj.R^2=0.97948

Fig. 60 shows the decrease of collected Raman signal at longer stand-off distances. Two effects cause the decrease of Raman intensity at longer distances. First, the reduction of collected Raman signal with increasing distance can be explained geometrically. The sample irradiates Raman scattered light in all directions, therefore, only a fraction of light can be collected via the telescope. This geometric effect causes a reduction with the square of the distance, as the imaginary surface of the spreading light sphere increases with the square of the radius, whereas the telescope collection area is constant. In Fig. 60 the regression for the geometric decrease is represented by a grey line, far off the data points. An intensity decay with the square of the distance can only be expected when the sample irradiates signal light equally in all directions. An opposite extreme would be a mirror which reflects the entire laser beam in only one single direction. If the returning light is detected along this direction, no distance dependency occurs. As can be seen in Fig. 60 the decrease in Raman intensity is neither distance independent nor does it change with the square of the distance. Hence this reason quadratic 1/r^2 was replaced by a variable parameter 1/r^p. With the gained flexibility the data can be fitted more accurately (adjusted $R^2=0.97948$) as the according thin black line in Fig. 60 illustrates. However, as the data fit was still not satisfactory an exponential term (exp(-k*distance)) was added to the equation to compensate for the losses due to absorption and scattering in the atmosphere. The laser power reaching the sample surface decreases along the distance, as air molecules, as well as airborne dust particles, scatter the photons elastically and inelastically. The Raman photons returning from the sample undergo many different scattering events in the atmosphere, depending on the size of the particles [87]. Whereas Mie scattering occurs on particles with a diameter similar to the incident light wavelength, Rayleigh scattering happens when the particles are smaller (i.e. air molecules). Scattering events on the sample as well as in the air can be the reason for the deviation from the signal intensity decrease with the square of the distance, described in literature [88]. This unexpected deviation (Fig. 60) can not be fully explained and would required further investigation.

With the resulting Equation 2 it is possible to explain the influence of stand-off distance on collected Raman signal sufficiently (bold black line in Fig. 60).

$$I(r) = H \frac{1}{r^p} \exp(-k \cdot r)$$

Equation 2: Raman intensity I at different stand-off distances r; k atmospheric loss, p distance proportionality, H includes laser power and substance specific Raman intensity

The fact that the velocity of light is a constant, it is expected that the remote distance where the sample is located would correspond directly to delay between the laser trigger pulse and when the camera gate is opened. This dependence is shown in Fig. 61 where the stand-off distance between measurement setup and the sample is plotted on the horizontal axis.

The vertical axis represents the ICCD delay between the trigger pulse from the laser and the opening of the camera gate.

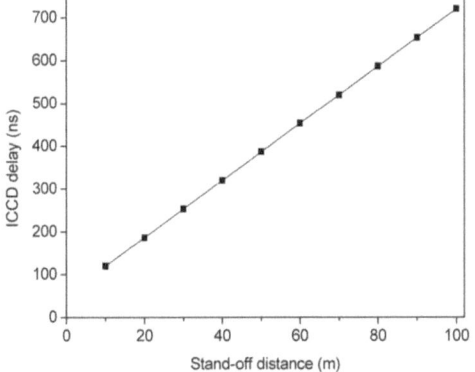

Fig. 61: Increase of ICCD camera delay with stand-off distance; ICCD delay=(53.3±0.1)+(6.670±0.002)*Stand-off distance; R^2=1

The slope of the linear relation between ICCD delay and stand-off distance would yield the speed of light. Taking into account that the light travels the distance between setup and sample twice, the slope in Fig. 61 should be 6.667 ns/m which is in good agreement with the determined value of 6.670±0.002 ns/m. Due to this well defined relation between the delay of the trigger pulse and the signal collection of the camera, time resolved measurements can be performed thus enabling not only chemical information but also location. This is the subject of recently accepted publication [89] and the content of chapter 5 of this book.

2.3.2 Stand-off distance influence on telescope depth of focus

For maximum signal collection efficiency in addition to the correct timing between laser trigger and ICCD delay the telescope needs to be focused at the correct sampling distance. This is achieved by moving the main mirror of the 6'' telescope to focus the collected light on the fibre optic fibre, which is attached at the back of the telescope. Depending on the stand-off distance between setup and sample, for each experiment the telescope has to be set for this particular distance. If the sample location deviates from this set distance the signal collection efficiency is reduced. The deviation range on either side of the set, optimal distance which still allows sufficient light collection is referred to as "depth of focus". An increase of the depth of focus with the square of the distance was described by Hirschfeld et al. [6]. To confirm this findings the depth of focus was determined at different sampling distances, ranging from 8 to 15 m.

To examine this effect for the used setup, spectra of air were recorded. Due to the homogeneous distribution of air no solid sample had to be placed at the various distances. Therefore, any influence by misalignment of a manually placed sample at each distance was avoided. Fig. 62 shows a representative air spectrum recorded with the telescope focus set to 10 m. For this experiment 600 green (532 nm) laser pulses with 153 mJ pulse energy were accumulated, using an ICCD gate of 500 ps.

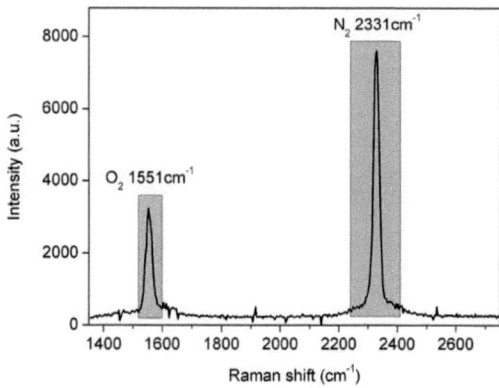

Fig. 62: 10 m-stand-off Raman spectrum of air, laser: 532 nm, 153 mJ/pulse, 600 pulses, 500 ps gate, 127 ns delay, grating 300 grooves/mm

Fig. 62 illustrates the band positions of oxygen O_2 (1551 cm^{-1}) and nitrogen N_2 (2331 cm^{-1}) in air. Furthermore, the integration ranges used to determine the collected Raman intensity are marked by rectangles. The oxygen band was integrated from 1525 to 1596 cm^{-1}. To determine the depth of focus, the telescope was focused at a particular distance and Raman spectra of air were recorded. To achieve a depth dependent signal the delay between laser pulse and the synchronised 500 ps camera gate was changed between each measured spectrum. To obtain the data shown at the bottom of Fig. 63 the telescope was focused at a distance of 8 m and the delay was changed from 107 to 123 ns, resulting in a signal change over delay time. To quantify this signal change the intensity of the recorded air oxygen band was integrated, as described above (Fig. 62). The change of integral values with the delay times was described by a Gaussian curve. The FWHM (Full Width at Half Maximum) of the Gaussian fit was determined as a measure of the depth of focus at the particular focal distance. The same procedure was carried out for 8 different focus distances, ranging from 8 to 15 m. For each of these telescope settings the delay was varied in steps of 1 ns. For focus distances ranging from 8 to 15 m, Fig. 63 summarises the change, of the intensity integrals of the collected air oxygen Raman band, with delay time.

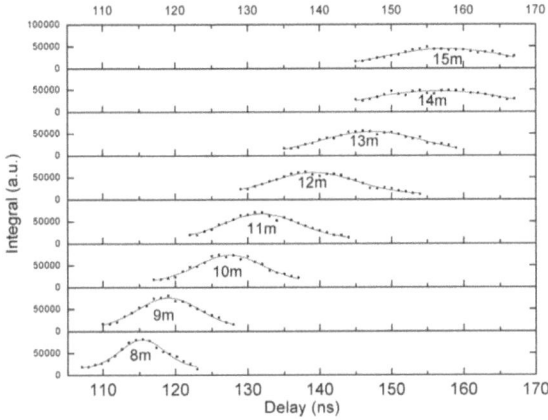

Fig. 63: Change of signal intensity with varying ICCD camera delay time relative to the laser pulse for 8 different measurement distances

From Fig. 63 can be seen that the optimum delay time increases with the focal distance, due to the time the light travels from the laser to the focal distance and back to the detection system. Furthermore, the absolute integral values decrease with the distance as the collectable Raman signal decreases with the stand-off distance. But most importantly for this particular experiment, is the broadening of the curves at longer distances. It means that the depth of focus increases when the telescope is focused at longer distances.

A quantitative representation, based on the widths of the Gaussian fits is given in Fig. 64 The FWHM in ns was converted into the depth of focus in metres, via the speed of light.

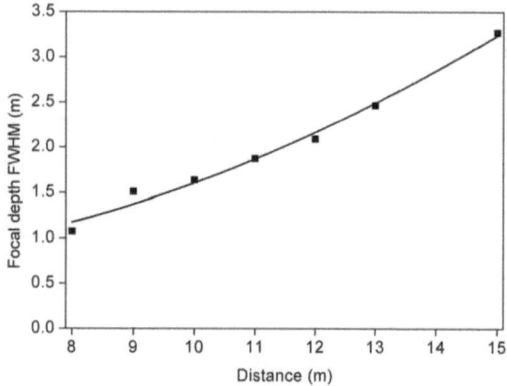

Fig. 64: Depth of focus of telescope; increase at higher focus distance; y=0.72-0.070*x+0.0159*x^2; R^2=0.982

As illustrated in Fig. 64 the depth of focus of the collecting telescope increases with the distance. This behaviour is observed since the probed air volume is constant in this experiment, defined by the 500 ps gate width of the ICCD camera. When the sampled gas volume is neither limited by the camera gate nor by the laser pulse length the signal should be distance independent [6], since the signal decrease with the distance is compensated by an increasing sample volume.

2.3.3 Spatial laser alignment

The alignment of laser beam and the optical collection system (the telescope) is of crucial importance to obtain good stand-off Raman spectra. In Fig. 65 the change of collection efficiency with spatial offset between laser excitation spot and telescope axis is shown. Raman spectra of a 1 mm thick polyethylene sheet were recorded at a stand-off distance of 9 m and the baseline corrected integral between 2500 and 3300 cm^{-1} was calculated.

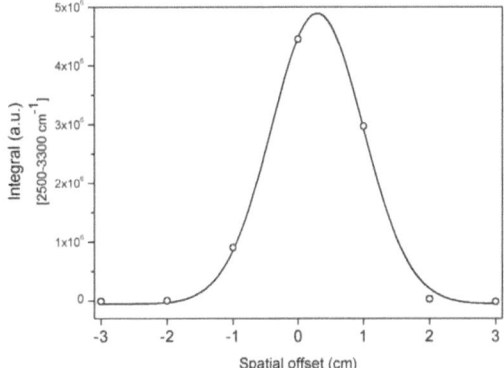

Fig. 65: Baseline-corrected integral of 9 m-stand-off Raman polyethylene bands between 2500 and 3300 cm^{-1} are drawn as a function of the laser offset from the telescope axis; a Gaussian curve was fitted through the measured data, leading to a FWHM (Full Width at Half Maximum) of 1.6 cm

In Fig. 65 the quick decrease of the integral – representing the collection efficiency – shows the importance of good alignment between laser beam and telescope. Appropriate misalignment can, however allow to determine the content inside of containers which are non-transparent to the human eye. This technique is detailed in two publications [27], [28] which are attached at the end of this book. Furthermore, this method called stand-off spatial offset Raman spectroscopy (stand-off SOSR) is discussed in a later chapter.

3 Applications of stand-off Raman spectroscopy

The previously described stand-off Raman system was used to identify chemical substances at remote distance of up to 100 m. The identification was not limited to pure substances but was also possible in the presence of interfering materials and on different background matrices. Furthermore, the quantification of liquids as well as solids is possible. Moreover, the feasibility of multivariate quantification of substance mixtures was shown. The results of qualitative and quantitative stand-off Raman spectroscopy are summarised in two publications [90], [14], which are attached at the end of this book. Here additional results are presented.

3.1 Qualitative analysis

To assess the applicability of stand-off Raman spectroscopy for explosives detection in real-world scenarios, their detection on different background materials and in the presence of interferents at a distance of 19 m was investigated. Furthermore, the measurement distance was extended to 100 m for the identification of eleven substances (explosives, precursors, chemicals and a war agent simulant). The excitation wavelength used was 532 nm along with the telescope suited for visible radiation.

3.1.1 Identification on backgrounds and with interferents

To meet the challenging circumstances in realistic scenarios combinations of different target analytes and interfering substances were defined by the OPTIX consortium. Furthermore, the laser pulse energy density was set to 70.7 mJ/cm^2 (pulse energy 20 mJ, beam diameter 6 mm) and the measurement time for the 20 m-stand-off Raman experiment was limited to 60 s. Analytes were placed on different background materials such as glass, LDPE (Low Density Poly Ethylene), aluminium, nylon as well as doors from three differently coloured cars.

Additionally, the nine analytes were mixed with 6 interferent materials; fuel oil, motor oil, soap, soil, NaCl and car wax. Tab. 7 summarises the measured combinations of analytes with background materials and interferents.

Tab. 7: Measurement matrix; 9 analytes (NaClO3 – EGDN) were measured pure, on 5 backgrounds (glass – car doors) and with 6 interferent materials (fuel oil – car wax)

	$NaClO_3$	$KClO_3$	NH_4NO_3	H_2O_2	PETN	TNT	DNT	RDX	EGDN
Glass	✓	✓	✓	✓	✓	✓	✓	✓	✓
LDPE	✓	✓	✓	✓	✓	✓	✓	✓	✓
Aluminium	✓	✓	✓	✓	✓	✓	✓	✓	✓
Nylon	✓	✓	✓	✓	✓	✓	✓	✓	✓
Car doors	✓	✓	✓	✓	✓	✓	✓	✓	✓
Fuel oil	✓	✓	✓	✓	✓	✓	✓	✓	✓
Motor oil	✓	✓	✓	✓	✓	✓	✓	✓	✓
Hand soap	✓	✓	✓	✓	✓	✓	✓	✓	✓
Soil	✓	✓	✓	✓	✓	✓	✓	✓	✓
NaCl	✓	✓	✓	✓	✓	✓	✓	✓	✓
Car wax	✓	✓	✓	✓	✓	✓	✓	✓	✓

Fig. 66 shows how the identification of sodium chlorate gets more difficult in the presence of interfering material, such as fluorescent motor oil.

Furthermore, identification is more challenging when an analyte has to be detected on a background material which gives Raman signals itself.

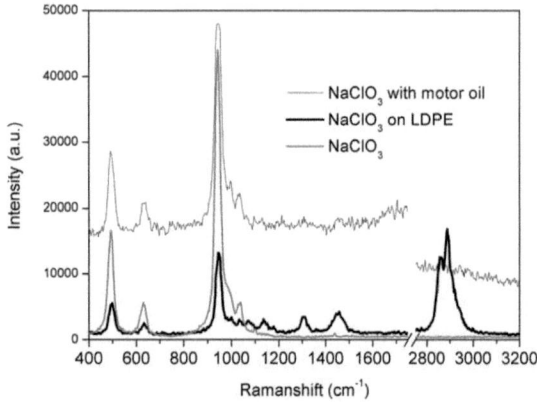

Fig. 66: 19 m-stand-off Raman spectra of NaClO₃ (100 laser pulses with 532 nm and 20 mJ): combination with fluorescent motor oil rises baseline; LDPE (Low Density Poly Ethylene) background leads to additional bands

Fig. 66 shows the result obtained when detecting $NaClO_3$ on different surfaces. For a more detailed overview the reader is referred to a publication by Zachhuber et al. [14] which is attached at the end of this book.

3.1.2 Identification of pure bulk substances at 100 m

Measurements conducted at 20 m distance at an earlier stage of the system development allowed identification, as long as fluorescence interference was not dominant. Due to further optimisation of the measurement setup during the project the spectral quality was improved significantly, allowing an extension of the 20 m distance defined in the OPTIX project to 100 m. To show the potential of stand-off Raman spectroscopy eleven substances were analysed at a stand-off distance of 100 m. Explosives (such as ANFO, TNT, RDX, PETN and HMX) used in industry as well as for military applications were analysed. Furthermore, potential explosive precursors such as sodium- and potassium chlorate ($NaClO_3$, $KClO_3$) as well as ammonium nitrate (NH_4NO_3) were also investigated, due to their general availability as weed killers or fertiliser. Triacetone triperoxide (TATP) was detected as a highly relevant substance used in Improvised Explosive Devices (IEDs).

In addition, spectra of DIMP (diisopropyl methylphosphonate) were recorded as this liquid is a chemical warfare agent simulant. Each substance (Fig. 67) was contained in a round glass vial (diameter 15 mm, height 20 mm) and placed 100 m from the apparatus. 600 laser pulses (532 nm) with a pulse energy of 228 mJ were accumulated.

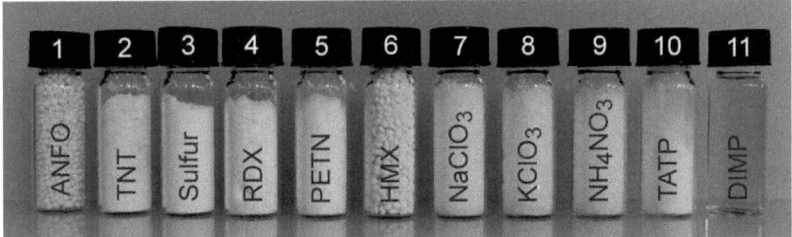

Fig. 67: Samples in glass bottles which were analysed at a distance of 100 m

Stand-off Raman spectra are shown in Fig. 68 and Fig. 69, together with the reference spectra, recorded on a confocal Raman microscope. The data between 1750 and 2750 cm^{-1} is not shown since this spectral range does not contain relevant information. When recording an extended spectral range it is necessary to move the grating of the spectrograph stepwise during the measurement. For these particular cases the spectra were recorded in two steps ranging from 400 to 1600 cm^{-1} and from 1600 to 3200 cm^{-1}, respectively.

APPLICATIONS OF STAND-OFF RAMAN SPECTROSCOPY

Fig. 68: Raman spectra; a: confocal microscope reference spectra; b: 100 m-stand-off spectra

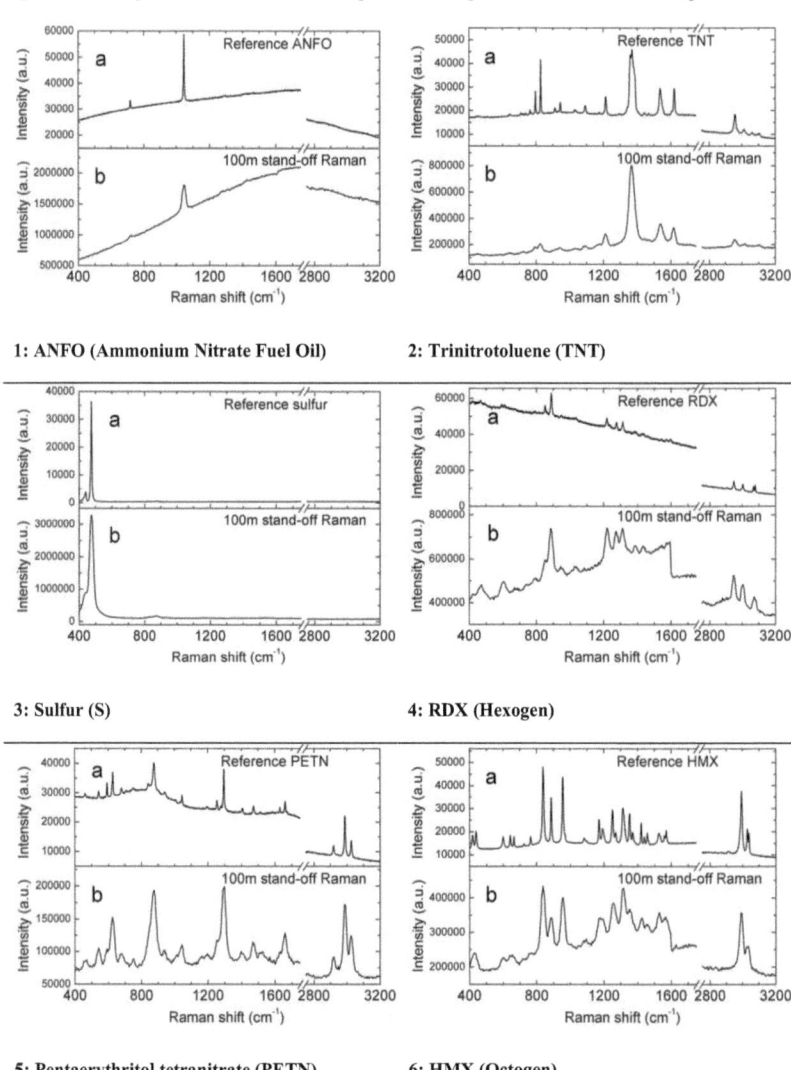

1: ANFO (Ammonium Nitrate Fuel Oil)

2: Trinitrotoluene (TNT)

3: Sulfur (S)

4: RDX (Hexogen)

5: Pentaerythritol tetranitrate (PETN)

6: HMX (Octogen)

Fig. 69: Raman spectra; a: confocal microscope reference spectra; b: 100 m-stand-off spectra

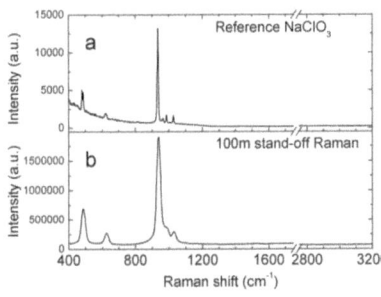

7: Sodium chlorate (NaClO₃)

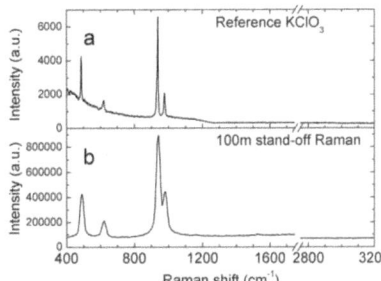

8: Potassium chlorate (KClO₃)

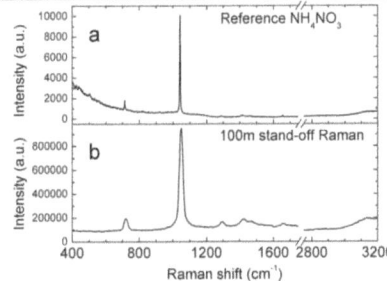

9: Ammonium nitrate (NH₄NO₃)

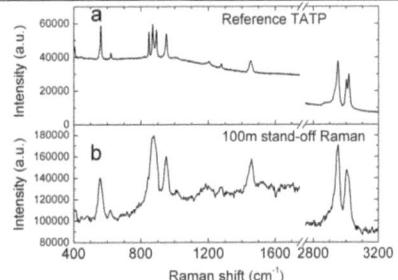

10: Triacetone triperoxide (TATP)

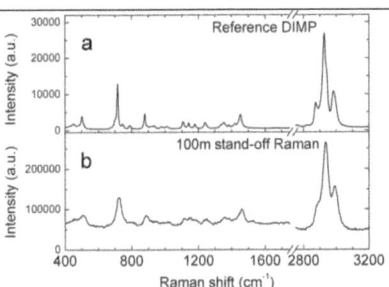

11: Diisopropyl methylphosphonate (DIMP)

3.2 Quantitative analysis

Quantitative analysis of stand-off Raman spectra was performed using both univariate and multivariate methods of data analysis. To correct for possible variation in instrumental parameters, the nitrogen band of ambient air was used as an internal standard. For the univariate method, stand-off Raman spectra obtained at a distance of 9 m on sodium chloride pellets containing varying amounts of ammonium nitrate (0-100%) were used, resulting in R^2 values of 0.992. For the multivariate quantification of ternary xylene mixtures (0-100%), stand-off spectra at a distance of 5 m were obtained. In Fig. 70a a stand-off Raman spectrum of a ternary mixture of o-, m- and p-xylene is illustrated. For comparison Fig. 70b shows 5 m-stand-off spectra of the pure isomers.

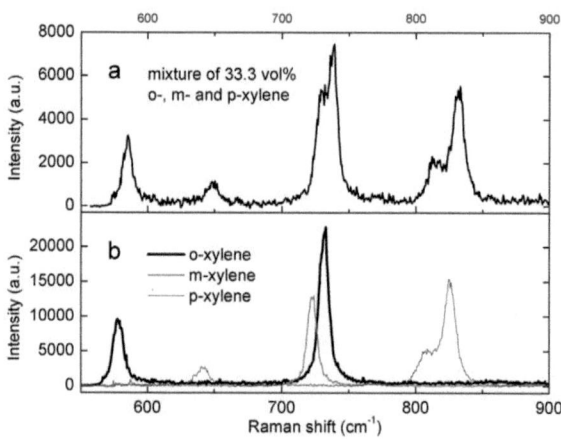

Fig. 70: 5 m-stand-off Raman spectra of xylene isomers; a: isomer mixture of o-, m- and p-xylene, 33.3% each; b: pure xylene isomers

The multivariate quantitative analysis yielded root mean square errors of prediction of 2.26%, 1.97% and 1.07% for o-, m- and p-xylene, respectively.

3.2.1 Trace detection and determination of limits of detection

For the determination of Limits Of Detection (LODs) the respective sample material was dissolved in an appropriate solvent and a defined volume of the solution was pipette on sample holders. The sample holders were aluminium plates with a dimple of 6 mm diameter in the middle. The dimple guaranteed sample deposition on a defined area which fitted the laser beam diameter of 6 mm. Inorganic salts such as $NaClO_3$, $KClO_3$ and NH_4NO_3 were solved in water whereas the explosives (DNT, TNT, PETN and EGDN) were dissolved in acetone. Each sample holder was placed at a distance of 19 m and 600 laser pulses with a pulse energy of 66 mJ and a wavelength of 532 nm were accumulated. The ICCD camera gate was set to 5 ns to collect the entire laser pulse with a pulse length of 4.4 ns.

3.2.1.1 Calculation of LOD

The calculation of the LOD is based on the baseline corrected integral of the most intense Raman band. The baseline was calculated as a linear regression through data points on either side of the integrated band. The boundaries of these two ranges are summarised in Tab. 8, together with the integration ranges and the determined LODs.

Tab. 8: Limit of detection (LOD) based on baseline corrected integrals; boundaries for integration and baseline calculation

Analyte	LOD (mg)	baseline range 1 (cm^{-1})	integration range (cm^{-1})	baseline range 2 (cm^{-1})
$KClO_3$	0.6410	778-869	917-1021	1050-1110
$NaClO_3$	0.6697	758-855	871-1007	1007-1085
NH_4NO_3	0.1358	902-1063	1017-1096	1096-1267
DNT	0.5317	1237-1266	1266-1442	1442-1482
TNT	0.6465	977-1029	1269-1496	1702-1833
PETN	0.5747	756-812	814-909	910-979
RDX	6.1454	729-833	834-934	935-1030

The baseline corrected integrals were plotted over the different analyte amounts. The obtained data sets showed a linear correlation of the integral and the amount of analyte. The slope q of this linear regression together with the residual standard deviation s allowed the calculation of the LOD, according to Equation 3 [91].

$$LOD = \frac{t \cdot s}{q} \sqrt{1 + \frac{1}{n} + \frac{c_\varnothing^2}{\sum (c_i - c_\varnothing)^2}}$$

Equation 3: LOD calculation via upper limit approach; t student factor, s residual standard deviation, q slop of calibration plot, n number of replicative measurements, $c_\varnothing$ mean concentration of calibration standards, c_i concentration of calibration standard

3.3 Stand-off Raman analysis via UV telescope

Compared to visible excitation, the use of UV lasers leads to increased Raman intensities. When reducing the excitation wavelength from 532 to 266 nm the Raman intensity is expected to increase by a factor of 16, since the intensity is proportional to 1/wavelength^4. Fig. 71 illustrates how the change in laser excitation wavelength additionally influences the resolution of the Raman spectrum.

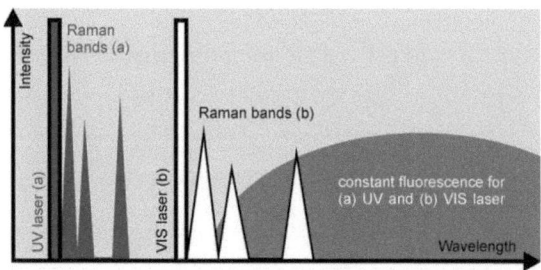

Fig. 71: Change of Raman spectrum with laser wavelength; excitation with UV laser (a) increases Raman intensity but reduces spectral resolution compared to visible (VIS) excitation (b); fluorescence is not necessarily influenced by the laser wavelength

In Fig. 71 the influence of a different laser excitation wavelength on the Raman and the fluorescence signal is shown. Whereas the Raman bands shift with the laser wavelengths, the fluorescence occurs at a constant wavelength. The selection of an appropriate laser wavelength results in Raman bands which do not coincide with disturbing fluorescence.

3.3.1 Excitation wavelength influence on Raman scattering

The change of the Raman signal with laser excitation wavelength is shown in this section for thirteen substances at a stand-off distance of 15 m. A pulse energy of 50 mJ was applied when exciting the samples with 532 nm. When using an excitation of 355 nm the pulse energy was set to 25 mJ and was further reduced to 12.5 mJ for a laser wavelength of 266 nm. The dispersion grating with 300 grooves/mm was sufficient when measuring with 532 or 355 nm. For 266 nm-measurements the grating was switched to 2400 grooves/mm, to maintain good spectral resolution. Reference spectra were recorded on a confocal Raman microscope (LabRAM system, Jobin-Yvon/Dilor, Lille, France). Raman scattering was excited by a HeNe laser at 632.8 nm.

The dispersive spectrometer was equipped with a grating of 600 lines/mm. The detector was a Peltier-cooled CCD detector (ISA, Edison, NJ).

3.3.1.1 Nylon

The recorded Raman data of a nylon sheet show the influence of laser excitation wavelength on the spectral response (Fig. 72).

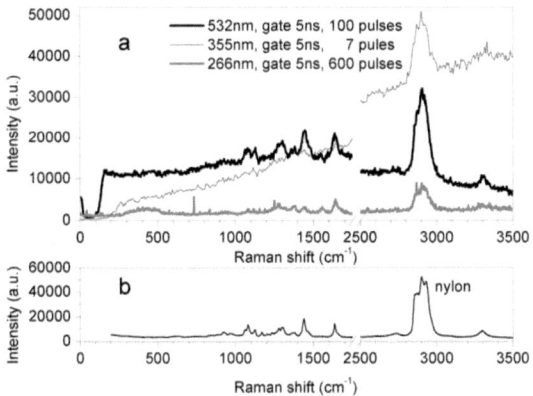

Fig. 72: 15 m-stand-off Raman spectra of nylon; a: measured at three different laser excitation wavelengths; b: reference spectrum measured via confocal Raman microscope

Whereas the spectrum (Fig. 72) collected after excitation with 532 nm shows sufficient Raman intensity on top of a moderately elevated baseline, excitation with 355 nm increases not only the Raman band at 2850 cm^{-1}, but also the fluorescence contribution. As a result, the bands between 1000 and 1700 cm^{-1} are not visible anymore. Further decrease in wavelength to 266 nm leads to a very flat baseline which reveals the latter bands despite their lower absolute intensity.

3.3.1.2 Low density polyethylene (LDPE)

The LDPE data in Fig. 73a shows that the use of shorter excitation wavelengths does not necessarily improve the spectral quality. At a Raman shift of 2850 cm^{-1} a similar band intensity is reached 20 times quicker with 355 nm-excitation than with 532 nm, despite the 50% lower pulse energy.

However, the 355 nm-laser excites fluorescence which results in an elevated background peaking at 1900 cm^{-1}. This fluorescence hides the LDPE bands between 1250 and 1500 cm^{-1} which are visible in the 532 nm spectrum.

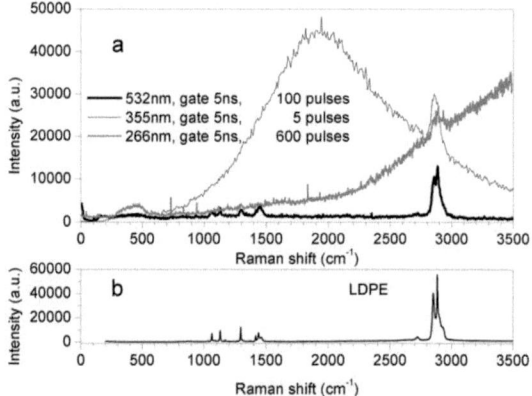

Fig. 73: 15 m-stand-off Raman spectra of LDPE; a: measured at three different laser excitation wavelengths; b: reference spectrum measured via confocal Raman microscope

Further reduction of the wavelength to 266 nm shifts the fluorescence to higher relative wavenumbers.

3.3.1.3 Sodium chlorate and potassium chlorate

The influence of the different laser wavelength for sodium chlorate and potassium chlorate (Fig. 74) shows how the collected Raman signal increases in the UV region.

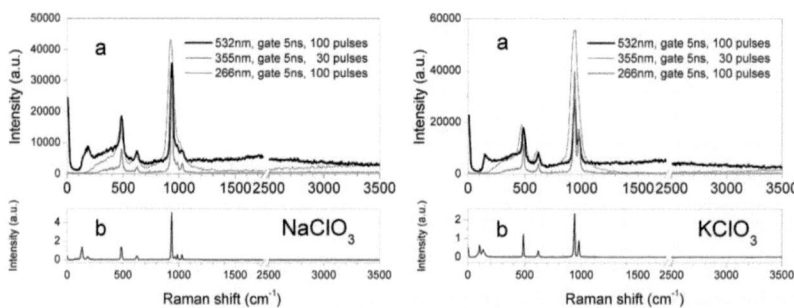

Fig. 74: 15 m-stand-off Raman spectra of NaClO$_3$ and KClO$_3$; a: measured at three different laser excitation wavelengths; b: reference spectrum measured via confocal Raman microscope

For both chlorates (Fig. 74) the laser pulse energy at 532 nm is twice as high as at 355 nm. Despite this fact and the shortage of the signal accumulation time to 30% the collected spectrum is more intense at 355 nm. Further reduction to 266 nm leads to spectra which are of similar intensity as the ones recorded at 532 nm. Bearing in mind that at 266 nm the pulse energy was reduced to 25% of the 532 nm-energy this is a considerable intensity gain.

3.3.1.4 Sulfur

Sulfur only shows a Raman response when excited at 532 nm, as can be seen in Fig. 75. The self absorption of the sample in the ultraviolet region, which leads to this behaviour is explained in more detail in a later section.

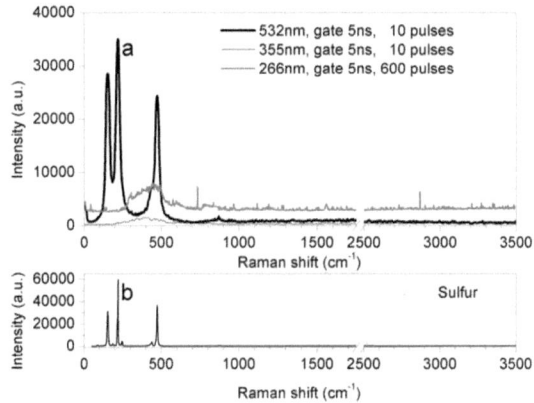

Fig. 75: 15 m-stand-off Raman spectra of sulfur; a: measured at three different laser excitation wavelengths; b: reference spectrum measured via confocal Raman microscope

3.3.1.5 Dinitrotoluene and Trinitrotoluene

For DNT as well as the structurally similar TNT the self absorption of the respective substance is the reason why only excitation at 532 nm results in Raman spectra (Fig. 76).

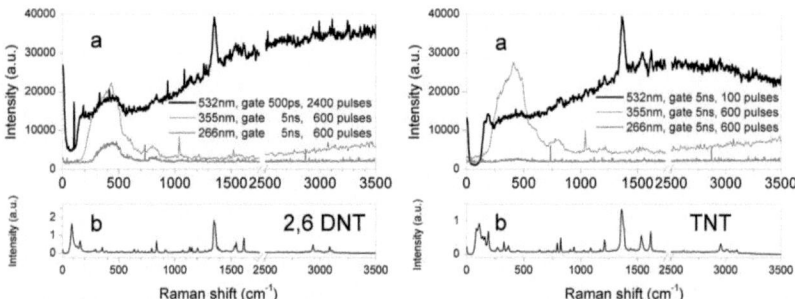

Fig. 76: 15 m-stand-off Raman spectra of DNT and TNT; a: measured at three different laser excitation wavelengths; b: reference spectrum measured via confocal Raman microscope

3.3.1.6 Absorption

The deviation from the expected Raman signal increase with decreasing laser excitation wavelength can be explained with the absorption of the tested solid materials. The diffuse reflectivity R of four powders was measured on a Lambda 750 UV/VIS spectrometer (Perkin Elmer). The absorption A (Fig. 77) was calculated according to the Kubelka-Munk equation: $A = (1-R/100)^2 / (2*R/100)$.

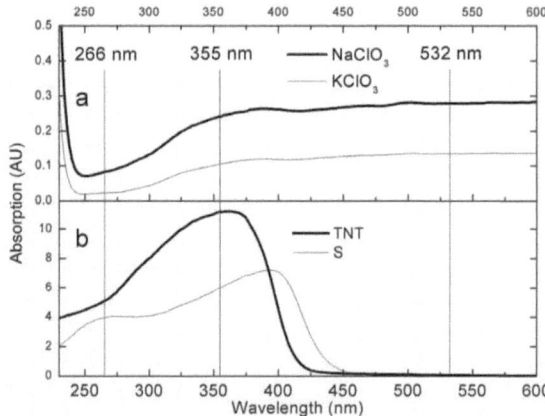

Fig. 77: UV/VIS absorption spectra; a: sodium chlorate ($NaClO_3$) and potassium chlorate ($KClO_3$); b: trinitrotoluene (TNT) and sulfur (S)

In Fig. 77a the absorption of the chlorates decreases slightly from 532 to 355 nm. A further reduction of the excitation wavelength to 266 nm significantly reduces the absorption. This is in accordance with the increased Raman signal which can be detected when exciting these samples at 355 and 266 nm, respectively. However, the absorption characteristics of TNT and sulfur (Fig. 77b) are completely different. Their low absorption at 532 nm shows a marked increase at 355 nm and a further wavelength reduction to 266 nm still leaves the absorption on a relatively high level. The self absorption in the UV region of the two latter materials is the reason why Raman excitation at these wavelengths does not lead to the expected Raman gain. To circumvent this obstacle, a tunable laser could be beneficial so that sample specific excitation laser wavelengths could be used. Also gaseous samples should have less absorption compared to solid samples [92].

Six additional substances (NH_4NO_3, H_2O_2, EGDN, RDX, PETN and DIMP) were measured using three different excitation laser wavelengths (532, 355 and 266 nm). The according Raman spectra are summarised in Fig. 78.

Fig. 78: Raman spectra via UV telescope; a: 15 m-stand-off spectra (532, 355 and 266 nm); b: reference spectra

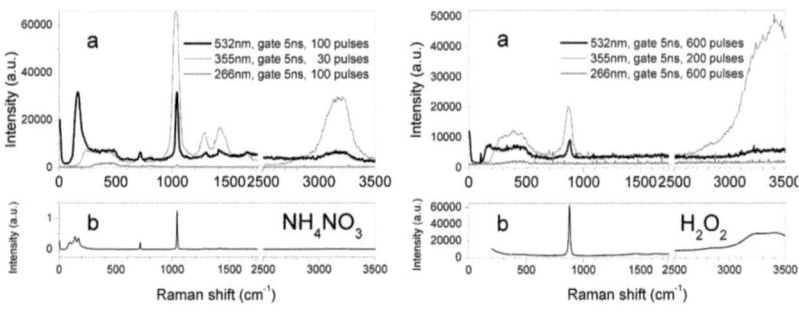

1: Ammonium nitrate (NH_4NO_3)
2: Hydrogen peroxide (H_2O_2)

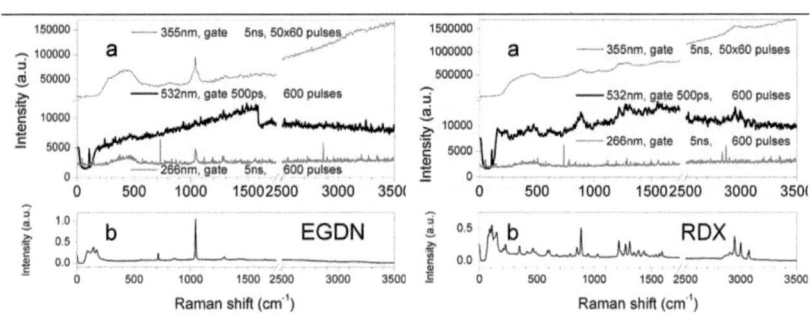

3: Ethylene glycol dinitrate (EGDN)
4: Hexogen (RDX)

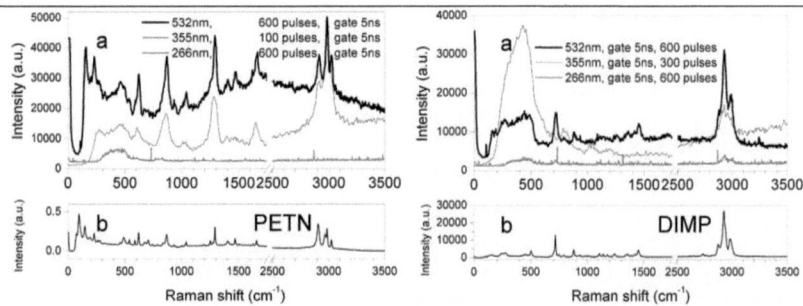

5: Pentaerythritol tetranitrate (PETN)
6: Diisopropyl methylphosphonate (DIMP)

4 Stand-off Spatial Offset Raman Scattering

4.1 Stand-off SORS principle

Spatial Offset Raman Scattering (SORS), developed by Matousek et al. [93] enables the detection of deeper layers of a sample by placing the collection optics at a position different (spatially offset) to the laser excitation spot on the sample surface. Stand-off detection (up to a distance of 40 m) of potentially hazardous materials concealed in containers or hidden behind opaque layers using a combination of stand-off Raman spectroscopy and SORS was introduced by Zachhuber et al. [27], [28]. This combination makes it possible to detect chemicals concealed in containers from a distance where a potential danger to the operator is reduced. Merging SORS with stand-off Raman spectroscopy, it is possible to detect liquids (i.e. isopropanol, hydrogen peroxide) as well as solid analytes (such as sodium chlorate) in opaque or white containers. This technique is not limited to qualitative analysis alone but can also be used for quantification. Furthermore, content of fluorescent containers, which is difficult, if not impossible for conventional (zero-offset) stand-off Raman spectroscopy measurements, can now be detected. The principle of stand-off SORS is in illustrated in Fig. 79.

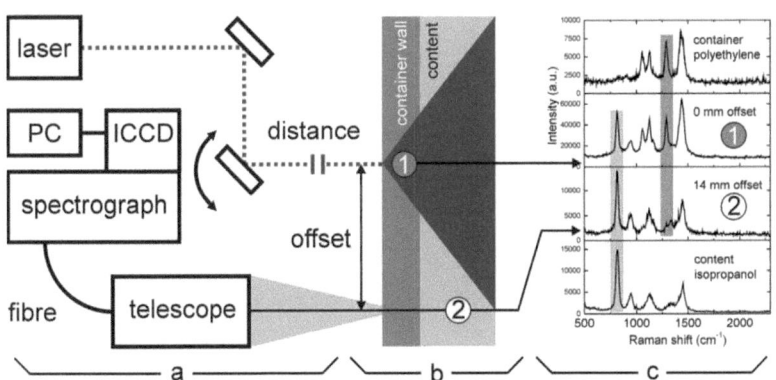

Fig. 79: Stand-off spatial offset Raman spectroscopy; a: setup, b: SORS principle; c: example spectra of isopropanol in a polyethylene container at 12 metres with different spatial offset

In Fig. 79b incident laser beam photons hitting a turbid container wall are strongly scattered, leading to a spatial spread of the laser beam in the probed material. Whereas in conventional Raman spectroscopy the inelastically scattered Raman signal is collected at the point (1) where the laser enters the material, with SORS, the detector is spatially moved away from this point (2). As a consequence the light reaching the detector originates primarily from deeper sample layers when laser spot and detector are spatially offset. The spectra depicted in Fig. 79c illustrate the change in relative band intensities when probing isopropanol in a turbid polyethylene (PE) bottle. At 0 mm offset (1) the Raman bands (highlighted with rectangles) of container (PE) and isopropanol content are both present. But as soon as laser and telescope are spatially offset (2) no PE signal is left, whereas the content signal is still clearly visible.

4.2 Line scan

To investigate the spatial resolution of the detection system, ten different analytes were filled into glass vials (15 mm diameter, filled height 20 mm). The glass vials were placed next to each other and a line scan across the ten samples at a stand-off distance of 100 m was performed. A motorised mirror mount was used to scan the laser spot (with a diameter of 6 mm) across the ten sample vials in steps of 4.5 mm resulting in 40 spectra shown in Fig. 80.

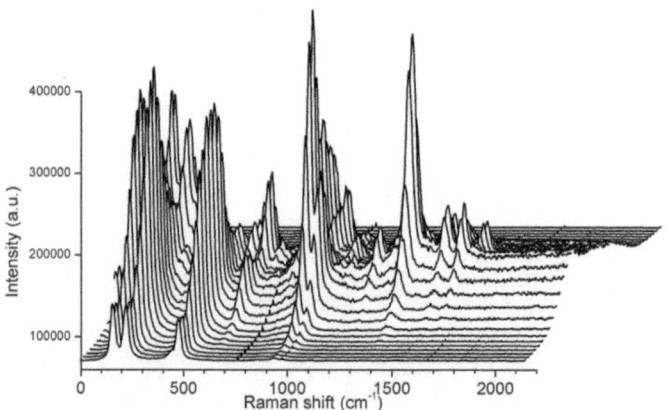

Fig. 80: 100 m-stand-off Raman line scan across ten different samples; the z axis represents the laser position on the samples; distance between consecutive spectra 4.5 mm

Fig. 80 shows the appearance of specific bands at certain laser positions.

For clarification, the spectra of ten different laser positions are shown in Fig. 81a, together with the reference spectra obtained via confocal microscope (Fig. 81b).

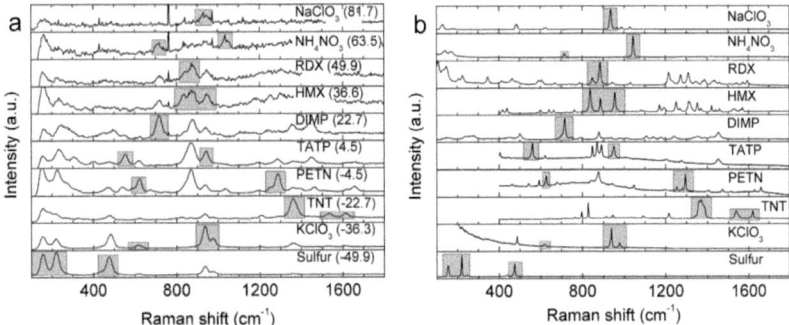

Fig. 81: Raman line scan; characteristic bands are marked with rectangles; a: 100 m-stand-off Raman spectra with determined sample position in mm in brackets; b: Raman reference spectra measured on a confocal microscope

In Fig. 81 characteristic bands for identification are emphasised with rectangles. The stand-off spectra (a) show bands of neighbouring substances. For example the band of $KClO_3$ at 938 cm^{-1} is present in the spectrum of sulfur and TNT. This can be explained with the movement of the laser spot on the sample due to atmospheric fluctuations. The low signal-to-noise ratio for RDX, NH_4NO_3 and $NaClO_3$ can be explained with the signal collection efficiency of the telescope, which decreases when the exciting laser spot is moved out of the optical telescope axis. This behaviour will be explained later in this section in more detail. For comparison Fig. 81b shows confocal Raman spectra of the same ten substances used for the line scan. To discriminate between HMX and RDX a more detailed look is needed, as the characteristic bands are located in overlapping spectral regions between 800 and 1000 cm^{-1}. RDX shows bands at and 851 and 888 cm^{-1} whereas the respective HMX bands are located at 837 and 886 cm^{-1}, allowing a substance differentiation.

The laser position of the most intense spectrum of each substance was used to determine the individual sample position. In Fig. 82 the found positions for the ten samples are indicated. The position with the highest signal collection efficiency, the telescope axis, was defined as zero.

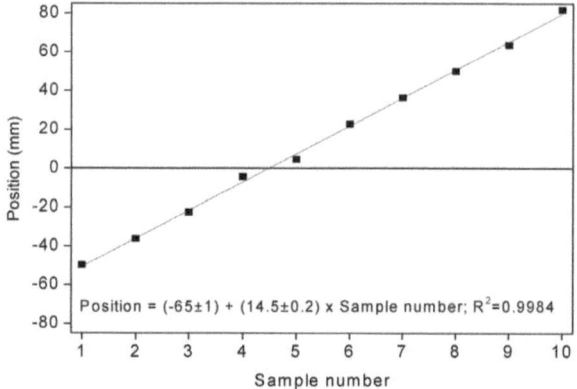

Fig. 82: Determined sample position of 10 measured samples and linear fit; position of highest collection efficiency was set to 0 mm

The linear correlation of sample number and position (Fig. 82) agrees well with the linear alignment of the sample glass vials. In Fig. 83 a line scan across a 1 mm-PP sheet at 100 m distance was conducted in order to establish the telescope Field Of View (FOV). The configuration and experimental parameters were identical to those used for the line sample scan, except for the signal accumulation time, which was reduced to 30 s per spectrum.

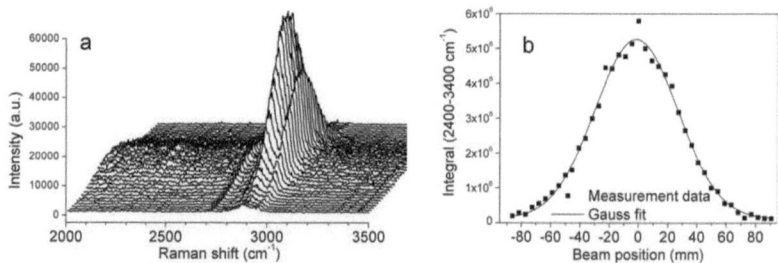

Fig. 83: 100 m-stand-off Raman line scan across a 1 mm-PP sheet, distance between spectra 4.5 mm; a: z-axis represents sampling position; b: baseline corrected integral (2400-3400 cm^{-1}) an Gaussian fit with FWHM 66.8 mm

In Fig. 83a 100 m-stand-off Raman spectra of a line scan across a 1 mm-PP sheet are shown. The 40 measurement spots were evenly distributed along a 177 mm scan line. Therefore, the distance between consecutive spots was 4.5 mm. Despite the homogeneous sample the signal intensity varies significantly. The band in the range of 2900 cm^{-1} in Fig. 83 shows a distinct intensity maximum when the laser aims at a sample position in the middle of the telescope axis. In order to evaluate the change of signal collection efficiency along the telescope diameter the baseline corrected integral from 2400 to 3400 cm^{-1} was calculated and is depicted in Fig. 83b. The PP band integral is represented by the black squares and a Gaussian fit was calculated to evaluate the change of collection efficiency with the offset between telescope axis and laser excitation spot. The determined FWHM of 66.8 mm agrees well with the theoretical FOV of 66.7 mm which was calculated [94] by multiplying the active diameter of the collection fibre optic bundle (1 mm) with the ratio of stand-off distance (100 m) and focal length of the telescope (1.5 m).

The experiment shows that, in addition to the analyte identification, it is possible to spatially resolve a substance distribution in a distance of 100 m.

4.3 Beam broadening with plastic thickness

The aim of the following experiment was the characterisation of the laser penetration in turbid media. In order to measure the broadening of the laser excitation zone with increasing sample penetration depth, turbid polypropylene (PP) sheets of different thickness (1, 2, 3, 4, 5, 6, 8 and 10 mm) were placed 12 m from the setup. The Raman signal excited by 48 laser pulses (532 nm, 20.3±0.3 mJ/pulse) was collected by a 6'' telescope. In order to visualise the spatial laser broadening the laser spot on the sample was scanned across the sample from left to right. In Fig. 84 spectra of a 10 mm-PP sheet are shown, measured with the laser on different spots on the sample.

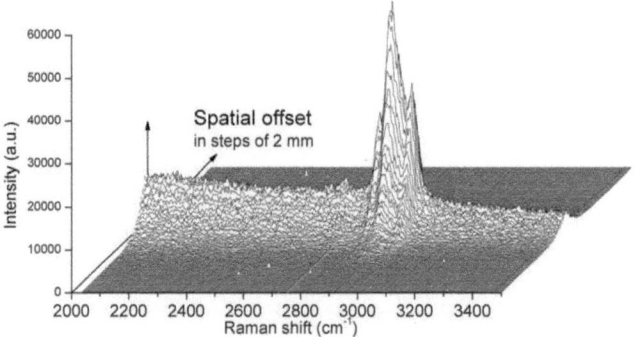

Fig. 84: Stand-off Raman spectra of 10 mm-PP sheet at 12 m distance; spatial offset between telescope axis and laser excitation spot changes along the third axis in steps of 2 mm; 2900 cm^{-1} band is maximum at 0 mm offset (when the laser coincides with the telescope axis)

The distance between the laser excitation spot on the sample surface and the optical axis of the collection optic is called spatial offset. This parameter changes along the third dimension in Fig. 84 in steps of 2 mm. Along this offset axis the distinct intensity maximum of the PP band at 2900 cm^{-1} marks the telescope axis, where the collection efficiency is best.

In order to evaluate the telescope Field Of View (FOV) baseline corrected integrals of PP spectra were calculated in the range from 2750 to 3050 cm^{-1}.

In Fig. 85 the change of the integral with spatial offset is shown for all eight analysed PP sheets of different thickness.

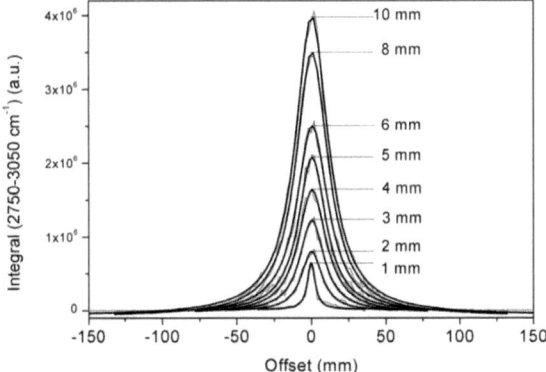

Fig. 85: Change of collected 12 m-stand-off Raman signal with spatial offset between telescope axis and laser illumination spot for eight PP sheets of different thickness; numbers in the diagram state the sample thickness; grey: original data, black: Lorenzian fits

The dependency of the collected signal intensity with changing offset was fitted by Lorenzian peaks, according to $y = y0 + (2*A/PI)*(FWHM/(4*(x-xc)^2 + FWHM^2))$. The individual Lorenz parameters for the different sized PP samples are summarised in Tab. 9.

Tab. 9: Lorenzian fit parameters for eight PP sheets of different thickness; y0 is height of the baseline, xc is the offset position with the highest intensity and FWHM represents the Full Width at Half Maximum. A states the peak area and Adj. R^2 is the adjusted correlation coefficient; H is the biggest, baseline corrected band height

thickness (mm)	y0 (a.u.)	xc (mm)	FWHM (mm)	A (a.u.*mm)	Adj. R^2	H (mm)
1	16575	-0.135	6.2	6.12E+06	0.96051	631730
2	-9923	0.153	16.8	2.15E+07	0.99328	813258
3	-22471	0.354	20.0	3.91E+07	0.99397	1248588
4	-43652	0.490	22.9	6.03E+07	0.99511	1672913
5	-31259	0.496	22.6	7.48E+07	0.99592	2108711
6	-54109	0.627	25.6	1.03E+08	0.99711	2557318
8	-78944	0.471	25.9	1.46E+08	0.99785	3590480
10	-67403	0.380	26.4	1.69E+08	0.99744	4069043

The maximum signal heights H of the PP samples (Tab. 9) were calculated according to H=2*A/(PI*FWHM). The linear relationship of H with the sample thickness is illustrated in Fig. 86.

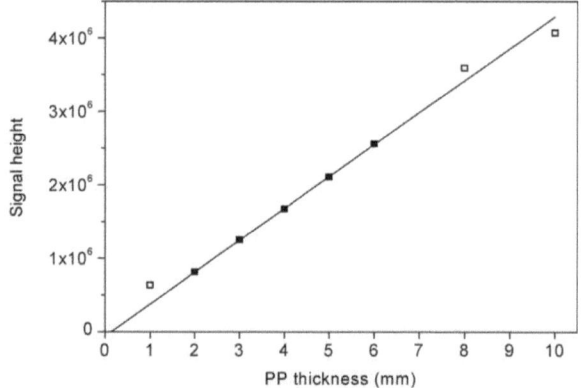

Fig. 86: Influence of PP thickness on maximum signal height H; linear regression based on data marked as black square, y=434828x-59140; white squares show little deviation from linear relationship

The determined linear regression of the signal heights H (Fig. 86) is based on the PP sheets with a thickness ranging from 2 to 6 mm. For these five samples the linear correlation coefficient R^2 of 0.99988 indicates a highly linear increase of signal height with growing sample thickness. For bigger sample thicknesses it is expected that the signal intensity reaches an upper boundary as the laser can not fully excite the whole sample anymore, due to absorption losses. Since neither the data point at 1 mm in Fig. 86 nor the data at 8 and 10 mm show significant deviation from the linear trend line, saturation is of no concern for PP sheets up to at least 10 mm thickness.

To specify the spatial laser broadening in the plastic, the Full Width at Half Maximum (FWHM) of the fitted Lorenzian peaks is of interest.

Fig. 87 shows the change of collected light with the spatial offset between laser spot and telescope axis.

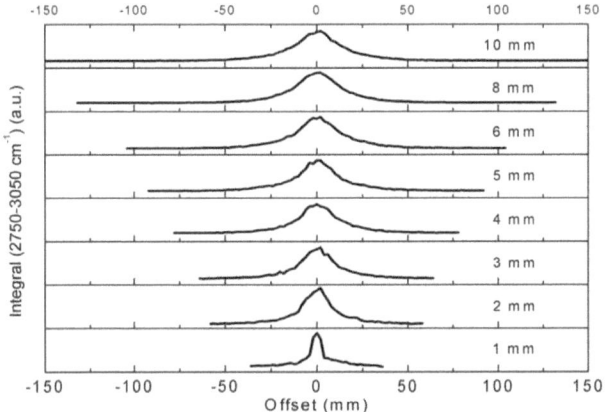

Fig. 87: Change of collected 12 m-stand-off Raman signal with spatial offset between telescope axis and laser illumination spot for eight PP sheets of different thickness; numbers in the diagram state the sample thickness

In Fig. 87 the increase of peak width with growing sample thickness can be seen. This broadening is a result of the diffuse scattering of the excitation laser in the turbid medium. For a more quantitative description the values for the FWHM (Tab. 9) are depicted in Fig. 88.

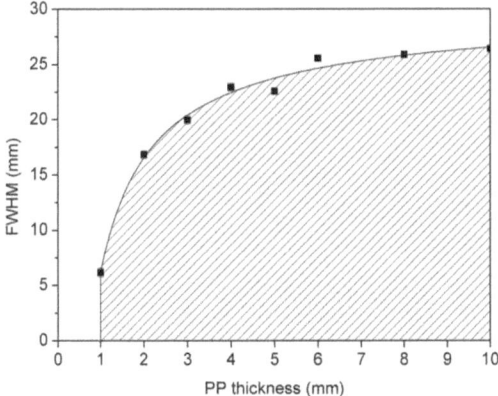

Fig. 88: Broadening of excited sample volume (FWHM) with increasing PP sample thickness; logistic fit y=A2+(A1-A2)/(1+(x/x0)^p) with A1=-2193, A2=30.2, x0=0.004, p=0.82 and R^2=0.97998; error bars (barely visible) represent standard errors

A logistic fit approximates the relationship between FWHM and sample thickness: w=30.2-2223.2/(1+(thickness/0.00408)^0.821). In Fig. 88 the broadening of the laser beam in the turbid PP sheets with increasing sample depth is visualised. Raman signal from a depth of 1.4 mm leaves the sample within a radius of 12.3 mm around the laser point of entry. During the next 3.8 mm this radius doubles, quadrupling the excited sample cross section at depths of 5.2 mm. With a diameter of 6 mm the employed laser beam was rather broad, reducing the accuracy of the absolute numbers. However, despite this limitation the significant spatial laser broadening can be shown.

This spatial laser broadening due to photon scattering in turbid media is the basis for further investigations using Spatial Offset Raman Spectroscopy (SORS) [95].

4.4 SORS in real world containers

In addition to the published stand-off SORS experiments [27] the capabilities of stand-off SORS were evaluated for the analysis of substances in containers present in real world scenarios: plastic flask for medicine, shampoo bottle, green glass bottle, woven plastic fabric and canvas shopping bag.

4.4.1 SORS in medicine flask

A white plastic bottle filled with $NaClO_3$ was analysed at a distance of 12 m. SORS spectra were recorded with an offset from -12 to +12 mm in steps of 0.5 mm.

In Fig. 89a a zero-offset spectrum (thin line) is shown together with a SORS spectrum of 5.5 mm offset (bold line).

When calculating the weighed difference of these two spectra data is obtained which shows only bands of the content or the bottle (Fig. 89b).

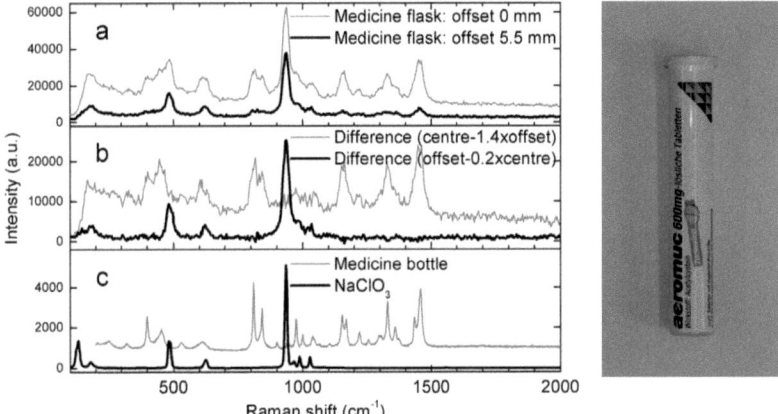

Fig. 89: Stand-off SORS of NaClO$_3$ in white medicine flask at 12 m; accumulation of 40 laser pulse with an energy of 10 mJ; a: 0 and 5.5 mm offset spectrum; b: weighed difference spectra; c: reference spectra

In Fig. 89b the thin line represents the difference between the zero-offset spectrum and 1.4 times the spectrum recorded at an offset of 5.5 mm. This calculated spectrum contains only bands of the bottle material. When calculating the difference of 5.5 mm-offset spectrum and 0.2 times the zero-offset spectrum (bold line in Fig. 89b) only the content bands of NaClO$_3$ are present. For Fig. 90a the recorded stand-off Raman band of the NaClO$_3$ content was integrated between 875 and 975 cm^{-1} (white circles) and the bands of the white plastic bottle were integrated from 1275 to 1575 cm^{-1} (black squares). With increasing offset the signal integrals of both materials decrease. As the integral of the NaClO$_3$ content decreases slower than the plastic bottle integral the ratio (Fig. 90b) increases at higher offsets.

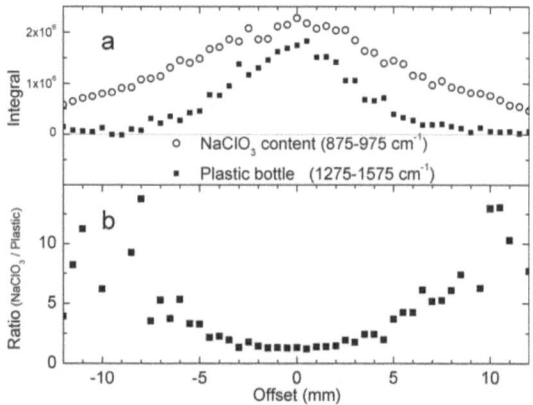

Fig. 90: 12 m-stand-off SORS of NaClO₃ in white medicine bottle; a: change of NaClO₃ and plastic bottle signal with offset; b: ratio of the two signal integrals vs. offset

4.4.2 SORS in Shampoo bottle

In Fig. 91a spectra of NaClO$_3$ in a white shampoo bottle at 0 and 5.0 mm offset are illustrated. The spectral difference is not very marked at first sight. But when calculating the difference spectra (Fig. 91b) it is possible to distinguish between the band of the content (bold line) and the bottle (thin line).

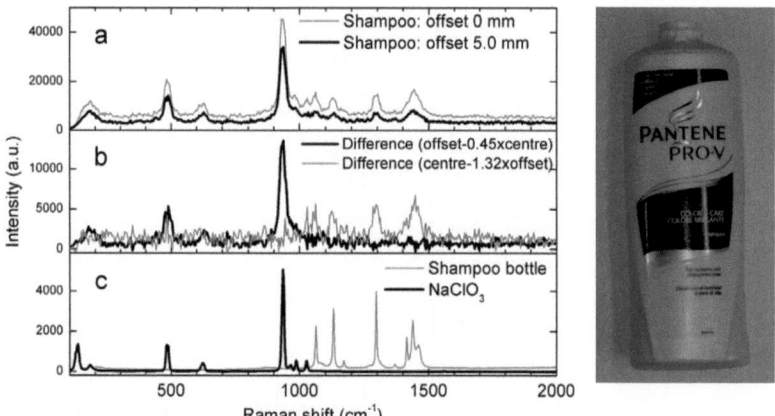

Fig. 91: Stand-off SORS of NaClO$_3$ in white shampoo bottle at 12 m; accumulation of 10 laser pulse with an energy of 10 mJ; a: 0 and 5.0 mm offset spectrum; b: weighed difference spectra; c: reference spectra

From this example it can be seen that the calculation of difference spectra is capable to reveal the origin of the individual signals in cases where the spectra are very similar to the human eye.

4.4.3 SORS in green glass bottle

A green glass bottle filled with NaClO$_3$ was analysed at a distance of 12 m. The spectra recorded at 0 or 6.0 mm offset, only vary in their absolute intensity, as can be seen in Fig. 92a.

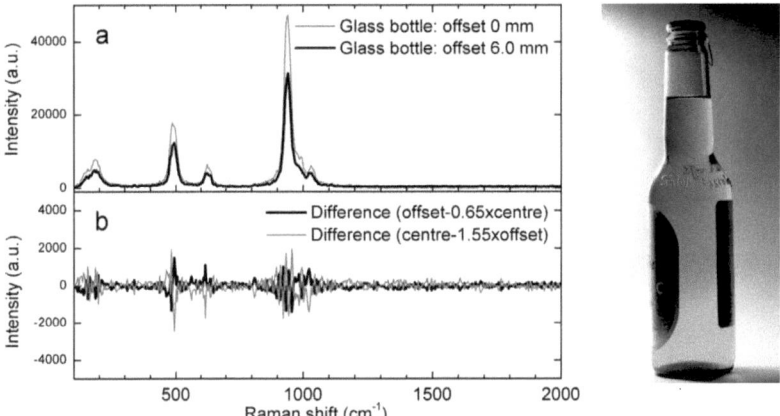

Fig. 92: Stand-off SORS of NaClO$_3$ in green glass bottle at 12 m; accumulation of 10 laser pulse with an energy of 10 mJ; a: 0 and 6.0 mm offset spectrum; b: weighed difference spectra

The difference spectra in Fig. 92b do not contain useful information. The reason for that is the spectral similarity at both offsets. The glass container does not contribute to the spectrum.

4.4.4 SORS in green woven nylon bag

A bag made of woven olive green plastic, as illustrated in Fig. 93 was filled with NaClO$_3$ and analysed from a distance of 12 m. This sample is challenging in several ways. First, the inhomogeneity of the woven structure does not scatter the laser as reproducible as a solid piece of plastic. Second, the coloured material causes intensive fluorescence hampering the content Raman signal.

Third, the strong laser absorption of the dark material reduces the amount of light entering the container content and furthermore allows only low laser powers when the bag should not be destroyed.

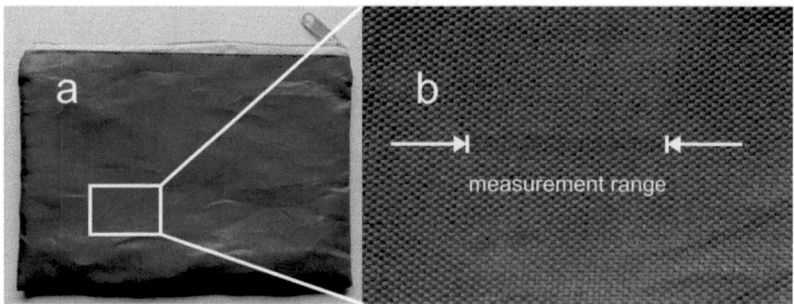

Fig. 93: Olive green bag made of woven nylon; a: overview; b: detail with marked measurement range

The spectrum at zero-offset (Fig. 94, thin line) shows extreme fluorescence. At Raman shifts greater than 1050 cm^{-1} the ICCD camera was even saturated. At an offset of 21.0 mm the fluorescence background is still very intense, but in addition bands of NaClO$_3$ can be identified.

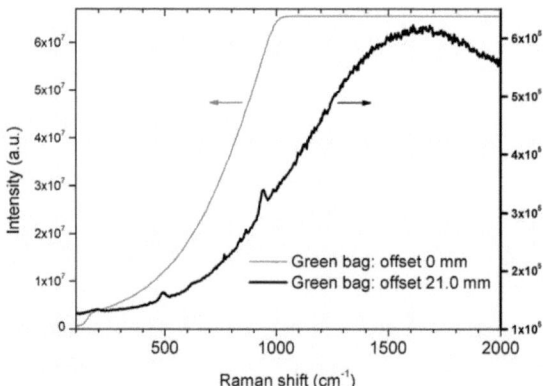

Fig. 94: Stand-off SORS of NaClO$_3$ in green woven nylon bag at 12 m; accumulation of 1000 laser pulses with an energy of 10 mJ

This challenging example shows how the range of measurable samples at stand-off distances can be expanded via stand-off SORS.

4.4.5 SORS in canvas shopping bag

For a colourless canvas shopping bag the stand-off distance was increased to 40 m. In Fig. 95 it can be seen that even without an offset a clear spectrum of the $NaClO_3$ content was obtained.

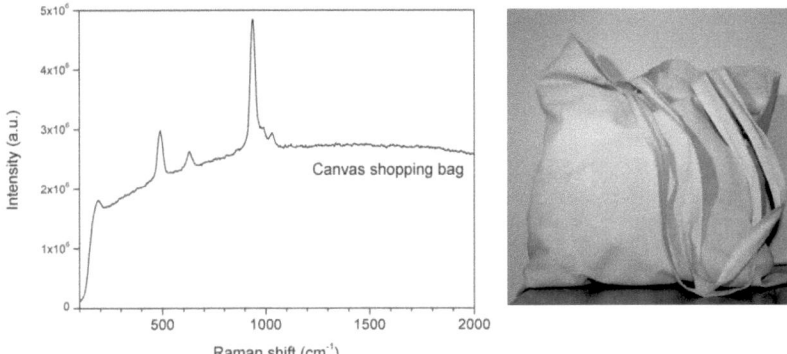

Fig. 95: Stand-off Raman spectrum of $NaClO_3$ in white canvas bag at 40 m; accumulation of 600 laser pulse with an energy of 20 mJ

In Fig. 95 the spectral analyte features on an elevated fluorescence background are strong enough to allow identification without SORS.

4.4.6 SORS at 40 metres

At 40 m stand-off distance a white HDPE (high density polyethylene) bottle (wall thickness 1.5 mm) filled with $NaClO_3$ was investigated. In Fig. 96a the zero-offset spectrum (thin line) looks quite similar to the spectrum recorded at an offset of 15.75 mm. Calculating the scaled difference spectra (Fig. 96b) the bands of the container (thin line) can be separated from the content band (bold line).

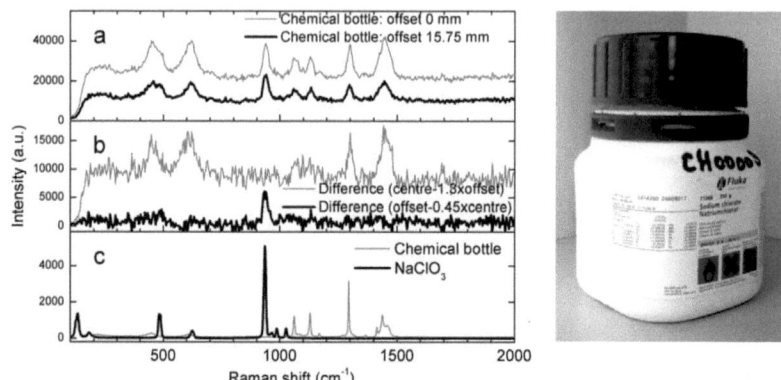

Fig. 96: Stand-off SORS of NaClO$_3$ in white chemical bottle at 40 m; accumulation of 200 laser pulses with an energy of 20 mJ; a: 0 and 15.75 mm offset spectrum; b: weighed difference spectra; c: reference spectra

Spectra at varying offsets were recorded and the bands of the plastic bottle (1225-1525 cm^{-1}) as well as the NaClO$_3$ content (875-975 cm^{-1}) integrated. The ratio of these two signal integrals are shown in Fig. 97.

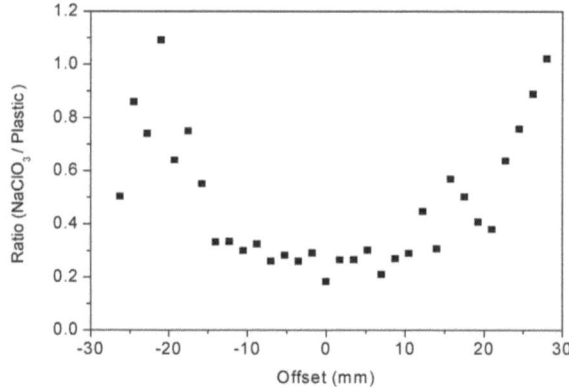

Fig. 97: 40 m-stand-off SORS of NaClO$_3$ in white chemical bottle; ratio of the two signal integrals vs. offset; integration ranges are for the NaClO$_3$ content 875-975 cm^{-1} and for the plastic bottle 1225-1525 cm^{-1}

From these results it can be seen that stand-off SORS can be extended to distances of tens of metres and its feasibility to detect the content in a variety of different containers present in real life scenarios was demonstrated.

5 Depth-resolved stand-off Raman spectroscopy

5.1 Measurement principle

The principle of time-resolved stand-off Raman spectroscopy is illustrated in Fig. 98 where two samples are placed at different distances from the detection system.

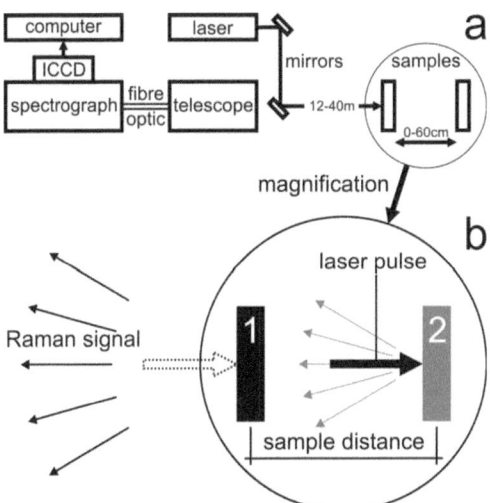

Fig. 98: Depth-resolved Raman spectroscopy; a: setup with two samples; b: sample distance leads to delay between Raman signal of individual substances

In Fig. 98 a laser pulse excites a substance thus yielding a specific Raman signal in sample 1. While the same laser pulse travels on to sample 2, the Raman signal of sample 1 already returns to the detection system. The introduced time delay between the two Raman signals permits calculation of the distance between the samples, using the speed of light.

5.2 Application of depth-resolved stand-off Raman spectroscopy

In a publication by Zachhuber et al. [89] attached at the end of this work, depth-resolved stand-off Raman spectroscopy was used to determine both, the identity and the position of substances relative to each other at remote distances (up to tens of metres). Spectral information of three xylene isomers, toluene and sodium chlorate was obtained at a distance of 12 m from the setup. Pairs and triplets of these samples were placed at varying distances (10-60 cm) relative to each other. Via the photon time of flight the distance between the individual samples was determined to an accuracy of 7% of the physically measured distance. Furthermore, at a distance of 40 m, time-resolved Raman depth profiling was used to detect sodium chlorate in a white plastic container which was non-transparent to the human eye. The combination of the ranging capabilities of Raman LIDAR (sample location usually determined using prior knowledge of the analyte of interest) with stand-off Raman spectroscopy (analyte detection at remote distances) provides the capability for depth-resolved identification of unknown substances and analysis of concealed content in distant objects. To achieve these results, a 532 nm-laser with a pulse length of 4.4 ns was synchronised to an intensified charge coupled device camera with a minimum gate width of 500 ps. For automated data analysis a multivariate curve resolution algorithm was employed.

In addition to the published data different substance combinations were investigated, the spectra of which are shown here. To obtain the spectra in Fig. 99 a quartz cuvette with toluene was placed 12.0 m from the setup, whereas a LDPE sheet was placed 0.55 m behind the toluene sample.

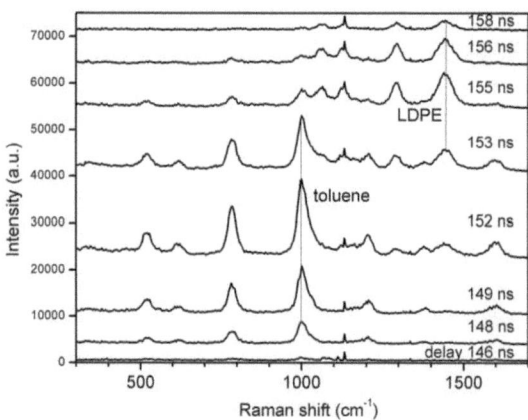

Fig. 99: Stand-off Raman spectra of toluene at 12.0 m and LDPE (low density polyethylene) at 12.55 m; different delay times in ns are stated next to the spectra

The different distance between the individual samples and the measurement apparatus is reflected by the different delay times (Fig. 99) between laser pulse and ICCD camera. Whereas the toluene signal intensity is maximum at a delay of 152 ns it peaks at

155.5 ns for the LDPE sheet. This time difference of 3.5 ns corresponds to 0.52 m which shows good agreement with the real distance of 0.55 m.

The result of a measured sample triplet is shown in Fig. 100. Toluene was placed 12.0 m from the setup, followed by o-xylene at 12.2 m and p-xylene at 12.4 m.

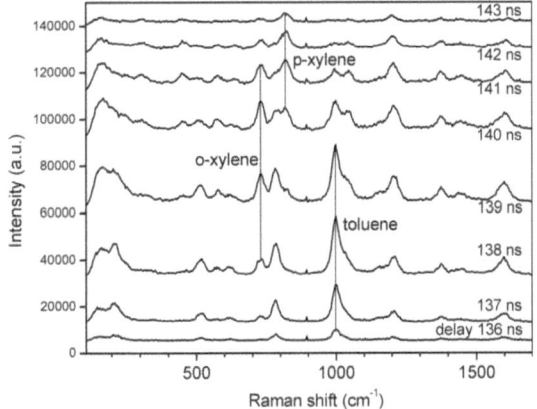

Fig. 100: **Stand-off Raman spectra of toluene, o-xylene and p-xylene placed at 12.0, 12.2 and 12.4 m; measured at different delay times (stated in the diagram)**

The delay time differences between the Raman intensity maxima for the individual samples allows to calculate their relative positions. The time delay of 1.7 ns between toluene and o-xylene leads to a calculated sample distance of 25.5 cm. 1.3 ns time difference results in a distance of 19.5 cm between o- and p-xylene.

The differences in the absolute delay time of the toluene maximum when comparing the two experiments described before (Fig. 99 and Fig. 100) were caused by a minor setup change. This change does not affect the relative differences between the individual samples.

6 Future plans for stand-off Raman spectroscopy

6.1 Three-dimensional stand-off Raman imaging

It was shown that it is possible to scan an excitation laser along one axis to achieve sample spectra from different positions. This concept can be extended to the second dimension, allowing chemical imaging [96]. In combination with depth-resolved stand-off Raman spectroscopy it is envisaged to record three-dimensional images of multi-substance dispersions. With 3D-stand-off Raman imaging the analysis in fluidised bed reactors could extend the knowledge about reaction dynamics since the distribution in space is accessible, rather than average values, typically obtained when removing samples from the process for analysis.

Furthermore, it was shown that depth-resolved stand-off Raman spectroscopy allows to identify substances concealed in containers which are non-transparent to the human eye. With a scanning laser beam this concept can be extended to multi-dimensional screening of wrapped material.

Another way to look into non-transparent opaque containers was found in stand-off spatial offset Raman scattering (SORS). How this concept can be used to obtain SORS images will be explained in the following section.

6.2 Stand-off SORS imaging

Two concepts to determine how a substance is distributed in a sample are presented. In both cases the lateral information is gained by scanning across the sample surface. In one case (direct SORS imaging) only the excitation laser is scanned while with the other method (dual scanning SORS) an additional movement of the collection optic (i.e. telescope) is necessary.

6.2.1 Dual Scanning SORS Imaging

As described above, this method is based on a scanning movement of the excitation laser as well as the telescope to obtain information about the lateral substance distribution in a sample. In Fig. 101 the dual scanning SORS principle is illustrated.

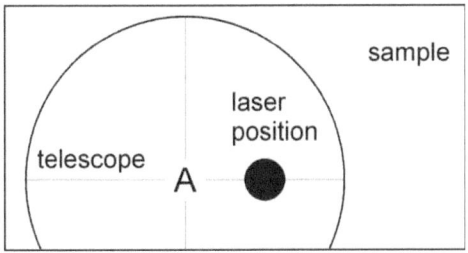

Fig. 101: Dual scanning SORS principle; the telescope field of view (big circle) is centred at position "A"; the excitation laser spot (black small circle) is spatially offset from the position "A"

The sample area which is observed by the telescope is indicated as big circle in Fig. 101. Spatially offset from the centre of the telescope field of view "A" the laser spot on the sample surface is illustrated as a smaller black circle. The spatial offset between telescope position and laser position can be varied (Fig. 102) to find the ideal offset [97] in a homogeneous sample or to gain depth dependent sample information.

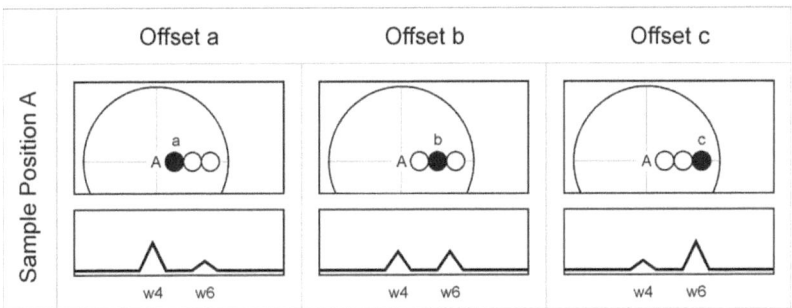

Fig. 102: Dual scanning SORS concept; the telescope remains centred on sample position "A" while the laser spot is moved to different spatial offsets (a, b, c); the spectra show bands at Raman shift "w4" and "w6" with changing relative intensities

Fig. 102 shows how the laser is scanned over the sample surface to realise different offsets (a, b, c) while the telescope remains at the sample position "A". In this way one sample position is analysed. In the shown example at an offset "a" the band at a Raman shift "w4" is more intense than the band at "w6". However, when increasing the offset to "b" and "c" the band ratio is reversed, leading to the conclusion that the substance associated with the band at "w6" is located at deeper layers than the other substance. In order to investigate a more extended sample area, the telescope has to be moved to different spots (Fig. 103).

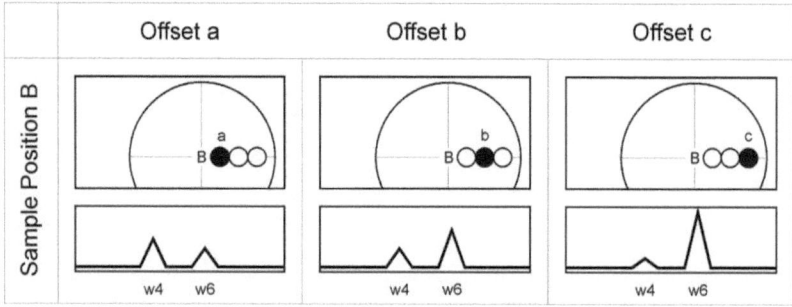

Fig. 103: Dual scanning SORS; the telescope field of view (big circle) was moved to sample position "B"; the laser (small black spot) is placed at varying spatial offsets (a, b, c); the recorded Raman spectra change with the offset

As illustrated in Fig. 103 the same procedure is repeated at the new sample position "B". The laser is moved to different positions relative to the telescope centre. The larger band ratio (w6/w4) at increased offset "b" and "c" shows that substance "w4" is located closer to the surface than "w6".

6.2.2 Direct SORS Imaging Concept

An alternative way to record SORS signals could facilitate a direct coupling imaging system described by Nordberg et al. [22]. Rather than focusing the collected light on a fibre optic cable the entire image is directly focused on a ICCD camera. Therefore, the lateral origin of the signals is not lost as the sample is imaged on the ICCD camera chip. In Fig. 104 a scheme of the camera chip is shown with the individual pixels.

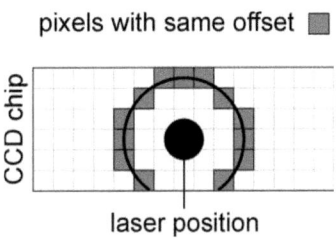

Fig. 104: ICCD camera chip with individual pixels; the black circle represents the laser position on the sample surface; grey squares are pixels at one constant spatial offset from the laser spot

The black circle indicates the position of the laser spot on the sample. With an expanded beam diameter this configuration was used to determine sulfur, ammonium nitrate, di- and trinitrotoluene [20]. The idea for SORS is, however to keep the laser diameter small and collect the signal at a position spatially offset from the point where the laser enters the sample. In Fig. 104 pixels at one constant offset from the laser are shown in grey. In Fig. 105 the laser is aimed at one particular position "A" on the sample, marked by the black circle.

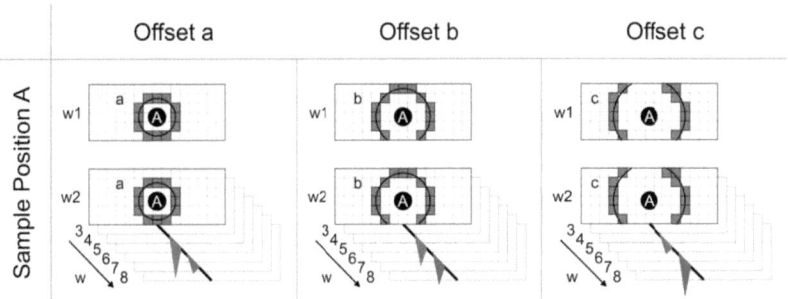

Fig. 105: Direct SORS imaging at sample position "A" (black circle); pixels of constant spatial offset (either a, b or c) from the laser spot are shown as grey squares; sequential collection of varying Raman shift (w1-w8) results in spectrum

To gain spectral information it is necessary to selectively determine the signal intensity at particular Raman shifts "w". This can be achieved by tunable filters [8], [20] sequentially collecting images at different Raman shifts. In Fig. 105 the spectrum collected at an offset "a" shows a band at a Raman shift "w4" and a smaller band at "w6". The convenience of this approach is that signals from various offsets (a, b, c) are collected simultaneously, as shown in Fig. 105. By selecting pixels at different offsets, the ideal offset [97] for a homogeneous sample can be found. Furthermore, using data at larger offset, sample information from deeper layers should be collected predominantly. In Fig. 105 the spectra change at varying offset. Whereas at an offset "a" the signal at a Raman shift "w4" is most intense, at an increased offset "c" the band at "w6" is more pronounced. This leads to the conclusion that the substance associated with the signal at "w6" is located at deeper layers of the sample. In this way it could be possible to gain depth dependent information. Especially, when combining SORS with time gated detection [98], [99]. To gain lateral information on the substance distribution, the laser can be scanned across the sample surface. In Fig. 106 the laser spot was moved to the position "B".

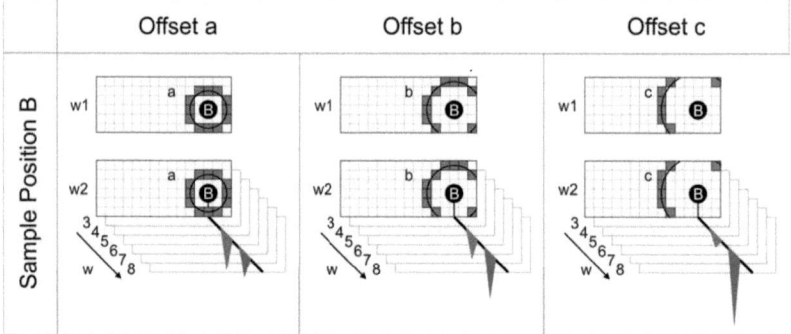

Fig. 106: Direct SORS imaging at sample position "B" (black circle); pixels of constant spatial offset (either a, b or c) from the laser spot are shown as grey squares; sequential collection of varying Raman shift (w1-w8) results in spectrum

Again, the signal at all offsets (a, b, c) is recorded simultaneously and the collected Raman shift tuned from "w1" to "w8" to gain the spectral information. Fig. 107 summarises the result of the direct SORS imaging concept

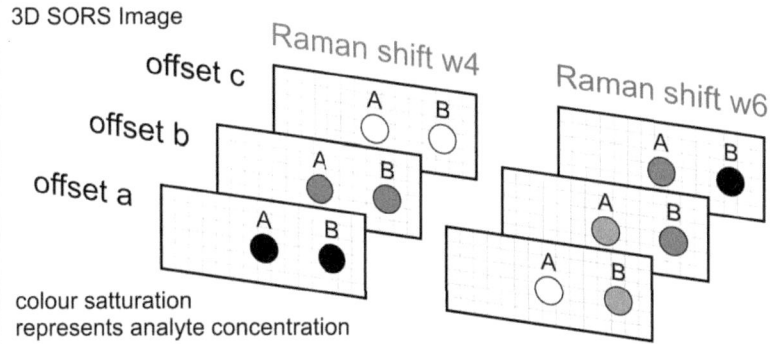

Fig. 107: Direct SORS 3D image for two substances associated with Raman shift "w4" and "w6", respectively; the substance concentration is represented by the saturation of the depicted circles; lateral information is achieved by moving the laser spot to different positions (A and B) on the sample surface; depths information could be gained using different spatial offsets (a, b, c)

In Fig. 107 the data recorded at the Raman shifts "w4" and "w6" are shown. The data from the different offsets (a, b, c) leads to depth information and the measurement at different sample spots (A, B) gives lateral information. In Fig. 107 the saturation of the shown circles represents the substance concentration. In the shown example the substance associated with the Raman signal at "w6" is most concentrated at deeper layers (offset c) at the position "B" but decreases when moving to the left (position A). In comparison, the other substance at "w4" is most concentrated at the sample surface (offset a) and shows a homogeneous lateral distribution (position A and B).

7 References

[1] M. Calvo Díez "OPTIX laser properties for Raman and LIBS" INDRA, Spain, Centro de Excelencia de Seguridad, pers. comm. 15/02/2012.

[2] M. Mordmüller "OPTIX laser properties for PLF-IR" Clausthal University of Technology, Germany, Laser Application Center, pers. comm. 16/02/2012.

[3] A. Pettersson, I. Johansson, S. Wallin, M. Nordberg and H. Östmark "Near Real-Time Standoff Detection of Explosives in a Realistic Outdoor Environment at 55 m Distance" Propellants, Explosives, Pyrotechnics, vol. 34, no. 4, pp. 297-306, 2009.

[4] C. Bauer, A. K. Sharma, U. Willer, J. Burgmeier, B. Braunschweig, W. Schade, S. Blaser, L. Hvozdara, A. Müller and G. Holl "Potentials and limits of mid-infrared laser spectroscopy for the detection of explosives" Applied Physics B, vol. 92, no. 3, pp. 327-333, 2008.

[5] J. Handke, F. Duschek, K. Gruenewald and C. Pargmann "Standoff detection applying laser-induced breakdown spectroscopy at the DLR laser test range" Proc. of SPIE vol. 8018, pp. 80180T-80180T-6, 2011.

[6] T. Hirschfeld "Range Independence of Signal in Variable Focus Remote Raman Spectrometry" Appl. Opt, vol. 13, no. 6, pp. 1435–1437, 1974.

[7] J. Cooney, K. Petri, and A. Salik "Measurements of high resolution atmospheric water-vapor profiles by use of a solar blind Raman lidar" Applied Optics, vol. 24, no. 1, pp. 104-108, 1985.

[8] J. C. Carter, J. Scaffidi, S. Burnett, B. Vasser, S. K. Sharma and S. M. Angel "Stand-off Raman detection using dispersive and tunable filter based systems" Spectrochimica acta. Part A, Molecular and biomolecular spectroscopy, vol. 61, no. 10, pp. 2288-98, 2005.

[9] S. K. Sharma, A. K. Misra, P. G. Lucey, S. M. Angel and C. P. McKay "Remote pulsed Raman spectroscopy of inorganic and organic materials to a radial distance of 100 meters" Applied spectroscopy, vol. 60, no. 8, pp. 871-6, 2006.

[10] S. M. Angel, N. R. Gomer, S. K. Sharma and C. McKay "Remote Raman Spectroscopy for Planetary Exploration: A Review" Applied Spectroscopy, vol. 66, no. 2, pp. 137-150, 2012.

[11] B. J. Bozlee, A. K. Misra, S. K. Sharma and M. Ingram "Remote Raman and fluorescence studies of mineral samples" Spectrochimica acta. Part A, Molecular and biomolecular spectroscopy, vol. 61, no. 10, pp. 2342-8, 2005.

REFERENCES

[12] P. Vandenabeele, K. Castro, M. Hargreaves, L. Moens, J. M. Madariaga and H. G. M. Edwards "Comparative study of mobile Raman instrumentation for art analysis" Analytica chimica acta, vol. 588, no. 1, pp. 108-16, 2007.

[13] J. C. Carter, S. M. Angel, M. Lawrence-Snyder, J. Scaffidi, R. E. Whipple and J. G. Reynolds "Standoff detection of high explosive materials at 50 meters in ambient light conditions using a small Raman instrument" Applied spectroscopy, vol. 59, no. 6, pp. 769-75, 2005.

[14] B. Zachhuber, G. Ramer, A. Hobro, E. t. H. Chrysostom and B. Lendl "Stand-off Raman spectroscopy: a powerful technique for qualitative and quantitative analysis of inorganic and organic compounds including explosives" Analytical and bioanalytical chemistry, vol. 400, no. 8, pp. 2439-47, 2011.

[15] A. J. Hobro and B. Lendl "Stand-off Raman spectroscopy" Trends in Analytical Chemistry, vol. 28, no. 11, pp. 1235-1242, 2009.

[16] S. Wallin, A. Pettersson, H. Östmark, A. Hobro, B. Zachhuber, B. Lendl, M. Mordmüller, C. Bauer, W. Schade, U. Willer, J. Laserna and P. Lucena "Standoff trace detection of explosives – a review" Proceedings of 12th Seminar on New Trends in Research of Energetic Materials, pp. 349-368, 2009.

[17] A. K. Misra, S. K. Sharma, D. E. Bates and T. E. Acosta "Compact standoff Raman system for detection of homemade explosives" Proc. of SPIE vol. 7665, pp. 76650U-76650U-11, 2010.

[18] J. Moros, J. A. Lorenzo, P. Lucena, L. M. Tobaria and J. J. Laserna "Simultaneous Raman spectroscopy-laser-induced breakdown spectroscopy for instant standoff analysis of explosives using a mobile integrated sensor platform" Analytical chemistry, vol. 82, no. 4, pp. 1389-400, 2010.

[19] J. Moros and J. J. Laserna "New Raman-Laser-Induced Breakdown Spectroscopy Identity of Explosives Using Parametric Data Fusion on an Integrated Sensing Platform" Analytical chemistry, vol. 83, no. 16, pp. 6275–6285, 2011.

[20] A. Pettersson, S. Wallin, H. Östmark, A. Ehlerding, I. Johansson, M. Nordberg, H. Ellis and A. Al-Khalili "Explosives standoff detection using Raman spectroscopy: from bulk towards trace detection" Proc. of SPIE vol. 7664, pp. 76641K-76641K-12, 2010.

[21] H. Östmark, M. Nordberg and T. E. Carlsson "Stand-off detection of explosives particles by multispectral imaging Raman spectroscopy" Applied optics, vol. 50, no. 28, pp. 5592-9, 2011.

[22] M. Nordberg, M. Åkeson, H. Ostmark and T. E. Carlsson "Stand-off detection of explosive particles by imaging Raman spectroscopy" Proc. of SPIE vol. 8017, pp. 80171B-80171B-7, 2011.

References

[23] R. L. Aggarwal, L. W. Farrar and D. L. Polla "Measurement of the absolute Raman scattering cross sections of sulfur and the standoff Raman detection of a 6-mm-thick sulfur specimen at 1500 m" Journal of Raman Spectroscopy, vol. 42, no. 3, pp. 461-464, 2011.

[24] F. Rull, A. Vegas, A. Sansano and P. Sobron "Analysis of Arctic ices by Remote Raman Spectroscopy" Spectrochimica acta. Part A, Molecular and biomolecular spectroscopy, vol. 80, no. 1, pp. 148-55, 2011.

[25] S. K. Sharma, A. K. Misra, S. M. Clegg, J. E. Barefield, R. C. Wiens, T. E. Acosta, D. E. Bates "Remote-Raman spectroscopic study of minerals under supercritical CO_2 relevant to Venus exploration" Spectrochimica acta. Part A, Molecular and biomolecular spectroscopy, vol. 80, no. 1, pp. 75-81, 2011.

[26] M. Åkeson, M. Nordberg, A. Ehlerding, L.-E. Nilsson, H. Östmark and P. Strömbeck "Picosecond laser pulses improves sensitivity in standoff explosive detection" Proc. of SPIE vol. 8017, pp. 80171C-80171C-8, 2011.

[27] B. Zachhuber, C. Gasser, E. t. H. Chrysostom and B. Lendl "Stand-off Spatial Offset Raman Spectroscopy for the detection of concealed content in distant objects" Analytical chemistry, vol. 83, pp. 9438-9442, 2011.

[28] B. Zachhuber, C. Gasser, A. J. Hobro, E. t. H. Chrysostom and B. Lendl "Stand-off spatial offset Raman spectroscopy: a distant look behind the scenes" Proc. of SPIE vol. 8189, pp. 818904-818904-8, 2011.

[29] O. Katz, A. Natan, Y. Silberberg and S. Rosenwaks "Standoff detection of trace amounts of solids by nonlinear Raman spectroscopy using shaped femtosecond pulses" Applied Physics Letters, vol. 92, no. 17, p. 171116, 2008.

[30] M. Dantus, H. Li, D. A. Harris, B. Xu, P. J. Wrzesinski and V. V. Lozovoy "Detection of chemicals at a standoff >10 m distance based on single-beam coherent anti-Stokes Raman scattering" Proc. of SPIE vol. 6954, pp. 69540P-69540P-5, 2008.

[31] A. Portnov, I. Bar and S. Rosenwaks "Highly sensitive standoff detection of explosives via backward coherent anti-Stokes Raman scattering" Applied Physics B, vol. 98, no. 2-3, pp. 529-535, 2009.

[32] M. T. Bremer, P. J. Wrzesinski, N. Butcher, V. V. Lozovoy and M. Dantus "Highly selective standoff detection and imaging of trace chemicals in a complex background using single-beam coherent anti-Stokes Raman scattering" Applied Physics Letters, vol. 99, no. 10, p. 101109, 2011.

[33] A. K. Misra, S. K. Sharma, T. E. Acosta and D. E. Bates "Compact remote Raman and LIBS system for detection of minerals, water, ices, and atmospheric gases for planetary exploration" Proc. of SPIE vol. 8032, pp. 80320Q-80320Q-12, 2011.

[34] R. Doyle and J. McNaboe "Stand-off detection of concealed improvised explosive devices (IEDs)" in Stand-Off Detection of Suicide Bombers and Mobile Subjects, pp. 77-87, 2006.

[35] D. C. Seward and T. Yukl, First International Symposium on Explosive Detection Technology, pp. 441-453, 1991.

[36] D. C. Seward and T. Yukl, Second Explosives Detection Technology Symposium & Aviation Security Technology Conference, pp. 162-169, 1996.

[37] J.-M. Thériault, E. Puckrin, J. Hancock, P. Lecavalier, C. J. Lepage and J. O. Jensen "Passive standoff detection of chemical warfare agents on surfaces" Applied optics, vol. 43, no. 31, pp. 5870-85, 2004.

[38] B. R. Cosofret, D. Konno, A. Faghfouri, H. S. Kindle, C. M. Gittins, M. L. Finson, T. E. Janov, M. J. Levreault, R. K. Miyashiro and W. J. Marinelli "Imaging sensor constellation for tomographic chemical cloud mapping" Applied optics, vol. 48, no. 10, pp. 1837-52, 2009.

[39] J. R. Castro-Suarez, L. C. Pacheco-Londono, W. Ortiz-Rivera, M. Velez-Reyes, M. Diem and S. P. Hernandez-Rivera "Open path FTIR detection of threat chemicals in air and on surfaces" Proc. of SPIE vol. 8012, pp. 801209-801209-13, 2011.

[40] A. Mukherjee, S. Von der Porten and C. K. N. Patel "Standoff detection of explosive substances at distances of up to 150 m" Applied optics, vol. 49, no. 11, pp. 2072-8, 2010.

[41] B. Hinkov, F. Fuchs, J. M. Kaster, Q. Yang, W. Bronner, R. Aidam and K. Köhler "Broad band tunable quantum cascade lasers for stand-off detection of explosives" Proc. of SPIE vol. 7484, pp. 748406-748406-13, 2009.

[42] R. D. Waterbury, A. Pal, D. K. Killinger, J. Rose, E. L. Dottery and G. Ontai "Standoff LIBS measurements of energetic materials using a 266nm excitation laser" Proc. of SPIE vol. 6954, pp. 695409-695409-5, 2008.

[43] J. L. Gottfried, F. C. De Lucia, C. a Munson and A. W. Miziolek "Laser-induced breakdown spectroscopy for detection of explosives residues: a review of recent advances, challenges, and future prospects" Analytical and bioanalytical chemistry, vol. 395, no. 2, pp. 283-300, 2009.

[44] D. Heflinger "Application of a unique scheme for remote detection of explosives" Optics Communications, vol. 204, no. 1-6, pp. 327-331, 2002.

[45] Y. Mou and J. W. Rabalais "Detection and identification of explosive particles in fingerprints using attenuated total reflection-Fourier transform infrared spectromicroscopy" Journal of forensic sciences, vol. 54, no. 4, pp. 846-50, 2009.

[46] M. Mordmueller, C. Bohling, A. John and W. Schade "Rapid test for the detection of hazardous microbiological material" Proc. of SPIE vol. 7484, pp. 74840F-74840F-10, 2009.

[47] V. Sivaprakasam, A. Huston, C. Scotto and J. Eversole "Multiple UV wavelength excitation and fluorescence of bioaerosols" Optics express, vol. 12, no. 19, pp. 4457-66, 2004.

[48] D. D. Tuschel, A. V. Mikhonin, B. E. Lemoff and S. a Asher "Deep ultraviolet resonance Raman excitation enables explosives detection" Applied spectroscopy, vol. 64, no. 4, pp. 425-32, 2010.

[49] G. Comanescu, C. K. Manka, J. Grun, S. Nikitin and D. Zabetakis "Identification of explosives with two-dimensional ultraviolet resonance Raman spectroscopy" Applied spectroscopy, vol. 62, no. 8, pp. 833-9, 2008.

[50] C. Carrillo, B. Lendl, B. M. Simonet and M. Valcárcel "Calix[8]arene coated CdSe/ZnS Quantum Dots as C60-nanosensor" Analytical chemistry, 2011.

[51] B. Zachhuber, C. Carrillo-Carrión, M. S. Suau and B. Lendl "Quantification of DNT isomers by capillary liquid chromatography using at-line SERS detection or multivariate analysis of SERS spectra of DNT isomer mixtures" Journal of Raman Spectroscopy, accepted 2011.

[52] F. Yan and T. Vodinh "Surface-enhanced Raman scattering detection of chemical and biological agents using a portable Raman integrated tunable sensor" Sensors and Actuators B: Chemical, vol. 121, no. 1, pp. 61-66, 2007.

[53] M. K. Khaing Oo, C.-F. Chang, Y. Sun and X. Fan "Rapid, sensitive DNT vapor detection with UV-assisted photo-chemically synthesized gold nanoparticle SERS substrates" The Analyst, vol. 136, no. 13, pp. 2811-7, 2011.

[54] G. Thomson and D. Batchelder "Development of a hand-held forensic-lidar for standoff detection of chemicals" Review of Scientific Instruments, vol. 73, no. 12, p. 4326, 2002.

[55] M. W. Todd, R. A.Provencal, T. G. Owano, B. A. Paldus, A. Kachanov, K. L. Vodopyanov, M. Hunter, S. L. Coy, J. I. Steinfeld and J. T. Arnold "Application of mid-infrared cavity-ringdown spectroscopy to trace explosives vapor detection using a broadly tunable (6-8 µm) optical parametric oscillator" Applied Physics B: Lasers and Optics, vol. 75, no. 2-3, pp. 367-376, 2002.

[56] U. Willer and W. Schade "Photonic sensor devices for explosive detection" Analytical and bioanalytical chemistry, vol. 395, no. 2, pp. 275-82, 2009.

[57] A. A. Kosterev, Y. A. Bakhirkin, R. F. Curl and F. K. Tittel "Quartz-enhanced photoacoustic spectroscopy" Optics letters, vol. 27, no. 21, pp. 1902-4, 2002.

[58] C. Bauer, U. Willer, R. Lewicki, A. Pohlkötter, A. Kosterev, D. Kosynkin, F. K. Tittel and W. Schade "A Mid-infrared QEPAS sensor device for TATP detection" Journal of Physics: Conference Series, vol. 157, p. 012002, 2009.

[59] C. W. Van Neste, L. R. Senesac and T. Thundat "Standoff spectroscopy of surface adsorbed chemicals" Analytical chemistry, vol. 81, no. 5, pp. 1952-6, 2009.

[60] M. E. Morales-Rodríguez, C. W. Van Neste, L. R. Senesac, S. M. Mahajan and T. Thundat "Ultra violet decomposition of surface adsorbed explosives investigated with infrared standoff spectroscopy" Sensors and Actuators B: Chemical, vol. 161, no. 1, pp. 961-966, 2012.

[61] Y. Song and R. G. Cooks "Atmospheric pressure ion/molecule reactions for the selective detection of nitroaromatic explosives using acetonitrile and air as reagents" Rapid communications in mass spectrometry RCM, vol. 20, no. 20, pp. 3130-8, 2006.

[62] Y. Takada, H. Nagano, M. Suga, Y. Hashimoto, M. Yamada, M. Sakairi, K. Kusumoto, T. Ota and J. Nakamura "Detection of Military Explosives by Atmospheric Pressure Chemical Ionization Mass Spectrometry with Counter-Flow Introduction" Propellants, Explosives, Pyrotechnics, vol. 27, no. 4, pp. 224-228, 2002.

[63] C. Mullen, A. Irwin, B. V. Pond, D. L. Huestis, M. J. Coggiola and H. Oser "Detection of explosives and explosives-related compounds by single photon laser ionization time-of-flight mass spectrometry" Analytical chemistry, vol. 78, no. 11, pp. 3807-14, 2006.

[64] G. Buttigieg "Characterization of the explosive triacetone triperoxide and detection by ion mobility spectrometry" Forensic Science International, vol. 135, no. 1, pp. 53-59, 2003.

[65] R. G. Ewing, D. A. Atkinson, G. A. Eiceman and G. J. Ewing "A critical review of ion mobility spectrometry for the detection of explosives and explosive related compounds" Talanta, vol. 54, no. 3, pp. 515-29, 2001.

[66] D. Gaurav, A. K. Malik and P. K. Rai "High-Performance Liquid Chromatographic Methods for the Analysis of Explosives" Critical Reviews in Analytical Chemistry, vol. 37, no. 4, pp. 227-268, 2007.

[67] R. Schulte-Ladbeck, A. Edelmann, G. Quintás, B. Lendl and U. Karst "Determination of peroxide-based explosives using liquid chromatography with on-line infrared detection" Analytical chemistry, vol. 78, no. 23, pp. 8150-5, 2006.

[68] J. Becanová, Z. Friedl and Z. Simek "Identification and determination of trinitrotoluenes and their degradation products using liquid chromatography-electrospray ionization mass spectrometry" International Journal of Mass Spectrometry, vol. 291, no. 3, pp. 133–139, 2010.

[69] Gaurav, V. Kaur, A. Kumar, A. K. Malik and P. K. Rai "SPME-HPLC: a new approach to the analysis of explosives" Journal of hazardous materials, vol. 147, no. 3, pp. 691-7, 2007.

[70] F.-F. Tian, J. Yu, J.-H. Hu, Y. Zhang, M.-X. Xie, Y. Liu, X.-F. Wang, H.-L. Liu and J. Han "Determination of emulsion explosives with Span-80 as emulsifier by gas chromatography-mass spectrometry" Journal of chromatography. A, vol. 1218, no. 22, pp. 3521-8, 2011.

[71] A. Halasz, C. Groom, E. Zhou, L. Paquet, C. Beaulieu, S. Deschamps, A. Corriveau, S. Thiboutot, G. Ampleman, C. Dubois, J. Hawari "Detection of explosives and their degradation products in soil environments" Journal of chromatography. A, vol. 963, no. 1-2, pp. 411-8, 2002.

[72] C. Sarazin, N. Delaunay, A. Varenne, C. Costanza, V. Eudes and P. Gareil "Capillary and Microchip Electrophoretic Analyses of Explosives and their Residues" Separation & Purification Reviews, vol. 39, no. 1-2, pp. 63-94, 2010.

[73] R. Schulte-Ladbeck, P. Kolla and U. Karst "A field test for the detection of peroxide-based explosives" Analyst, vol. 127, pp. 1152-1154, 2002.

[74] M. Amani, Y. Chu, K. L. Waterman, C. M. Hurley, M. J. Platek and O. J. Gregory "Detection of triacetone triperoxide (TATP) using a thermodynamic based gas sensor" Sensors and Actuators B: Chemical, vol. 162, no. 1, pp. 7-13, 2012.

[75] D. Lubczyk, C. Siering, J. Lörgen, Z. B. Shifrina, K. Müllen and S. R. Waldvogel "Simple and sensitive online detection of triacetone triperoxide explosive" Sensors and Actuators B: Chemical, vol. 143, no. 2, pp. 561-566, 2010.

[76] P. K. Sekhar, E. L. Brosha, R. Mukundan, K. L. Linker, C. Brusseau and F. H. Garzon "Trace detection and discrimination of explosives using electrochemical potentiometric gas sensors" Journal of hazardous materials, vol. 190, no. 1-3, pp. 125-32, 2011.

[77] R. Orghici, P. Lützow, J. Burgmeier, J. Koch, H. Heidrich, W. Schade, N. Welschoff and S. Waldvogel "A Microring Resonator Sensor for Sensitive Detection of 1,3,5-Trinitrotoluene (TNT)" Sensors, vol. 10, no. 7, pp. 6788-6795, 2010.

[78] F. Chu and J. Yang "Coil-shaped plastic optical fiber sensor heads for fluorescence quenching based TNT sensing" Sensors and Actuators A: Physical, vol. 175, pp. 43-46, 2012.

REFERENCES

[79] J. Clarkson "A theoretical study of the structure and vibrations of 2,4,6-trinitrotolune" Journal of Molecular Structure, vol. 648, no. 3, pp. 203-214, 2003.

[80] Y. A. Gruzdkov and Y. M. Gupta "Vibrational Properties and Structure of Pentaerythritol Tetranitrate" The Journal of Physical Chemistry A, vol. 105, no. 25, pp. 6197-6202, 2001.

[81] B. Brauer, F. Dubnikova, Y. Zeiri, R. Kosloff and R. B. Gerber "Vibrational spectroscopy of triacetone triperoxide (TATP): anharmonic fundamentals, overtones and combination bands" Spectrochimica acta. Part A, Molecular and biomolecular spectroscopy, vol. 71, no. 4, pp. 1438-45, 2008.

[82] J. Köhler and R. Meyer "Explosivstoffe" Wiley Online Library, ISBN 3-527-28327-7, 1998.

[83] W. T. Silfvast and W. B. Bridges "Laser fundamentals" Cambridge University Press Cambridge, England, ISBN 0-521-55617-1, 1996.

[84] E. Smith and D. Geoffrey "Modern Raman Spectroscopy - a practical approach" ISBN 978-0-471-49794-3, 2005.

[85] A. K. Misra, S. K. Sharma, C. H. Chio, P. G. Lucey and B. Lienert "Pulsed remote Raman system for daytime measurements of mineral spectra" Spectrochimica acta. Part A, Molecular and biomolecular spectroscopy, vol. 61, no. 10, pp. 2281-7, 2005.

[86] E. C. Cull "Standoff raman spectroscopy system for remote chemical detection" Proc. of SPIE vol. 5994, no. 2005, pp. 59940H-59940H-8, 2005.

[87] D. Sinclair and V. K. La Mer "Light Scattering as a Measure of Particle Size in Aerosols. The Production of Monodisperse Aerosols" Chemical Reviews, vol. 44, no. 2, pp. 245-267, 1949.

[88] L. C. Pacheco-Londoño, W. Ortiz-Rivera, O. M. Primera-Pedrozo and S. P. Hernández-Rivera "Vibrational spectroscopy standoff detection of explosives" Analytical and bioanalytical chemistry, vol. 395, no. 2, pp. 323-35, 2009.

[89] B. Zachhuber, C. Gasser, G. Ramer, E. t. H. Chrysostom and B. Lendl "Depth profiling for the identification of unknown substances and concealed content at remote distances using time resolved stand-off Raman spectroscopy" Applied spectroscopy, accepted.

[90] B. Zachhuber, G. Ramer, A. J. Hobro, and B. Lendl "Stand-off Raman spectroscopy of explosives" Proc. of SPIE vol. 7838, pp. 78380F-78380F-10, 2010.

[91] J. Mocak, A. M. Bond, S. Mitchell and G. Scollary "A statistical overview of standard (IUPAC and ACS) and new procedures for determining the limits of detection and quantification: application to voltammetric and stripping techniques" Pure and Applied Chemistry, vol. 69, no. 2, pp. 297-328, 1997.

[92] W. Kiefer "Self absorption for UV Raman" University of Würzburg, Germany, Institute for physical chemistry, pers. comm. 20/01/2012.

[93] P. Matousek, I. P. Clark, E. R. C. Draper, M. D. Morris, A. E. Goodship, N. Everall, M. Towrie, W. F. Finney and A. W. Parker "Subsurface Probing in Diffusely Scattering Media Using Spatially Offset Raman Spectroscopy" Applied Spectroscopy, vol. 59, no. 4, pp. 393-400, 2005.

[94] M. Wu, M. Ray, K. H. Fung, M. W. Ruckman, D. Harder and A. J. Sedlacek "Stand-off Detection of Chemicals by UV Raman Spectroscopy" Applied Spectroscopy, vol. 54, no. 6, pp. 800-806, 2000.

[95] P. Matousek, M. D. Morris, N. Everall, I. P. Clark, M. Towrie, E. Draper, A. Goodship and A. W. Parker "Numerical simulations of subsurface probing in diffusely scattering media using spatially offset Raman spectroscopy" Applied spectroscopy, vol. 59, no. 12, pp. 1485-92, 2005.

[96] J. N. Porter, C. E. Helsley, S. K. Sharma, A. K. Misra, D. E. Bates and B. R. Lienert "Two-dimensional standoff Raman measurements of distant samples" Journal of Raman Spectroscopy, accepted 2011.

[97] J. R. Maher and A. J. Berger "Determination of ideal offset for spatially offset Raman spectroscopy" Applied spectroscopy, vol. 64, no. 1, pp. 61-5, 2010.

[98] I. E. Iping Petterson, P. Dvořák, J. B. Buijs, C. Gooijer and F. Ariese "Time-resolved spatially offset Raman spectroscopy for depth analysis of diffusely scattering layers" The Analyst, vol. 135, no. 12, pp. 3255-9, 2010.

[99] I. E. Iping Petterson, M. López-López, C. García-Ruiz, C. Gooijer, J. B. Buijs and F. Ariese "Noninvasive Detection of Concealed Explosives: Depth Profiling through Opaque Plastics by Time-Resolved Raman Spectroscopy" Analytical chemistry, vol. 83, no. 22, pp. 8517-23, 2011.

Publication 1

S. Wallin, A. Pettersson, H. Östmark, A.J. Hobro, <u>B. Zachhuber</u>, B. Lendl, M. Mordmüller, C. Bauer, W. Schade, U. Willer, J. Laserna and P. Lucena

Standoff trace detection of explosives – a review

Proceedings of 12th Seminar on New Trends in Research of Energetic Materials (2009)
ISBN 978-80-7395-156, 349-368

STANDOFF TRACE DETECTION OF EXPLOSIVES – A REVIEW

Sara Wallin*, Anna Pettersson*, Henric Östmark*,
Alison Hobro**, Bernhard Zachhuber**, Bernhard Lendl**, Mario Mordmüller***,
Christoph Bauer***, Wolfgang Schade***, Ulrike Willer***,
Javier Laserna and Patricia Lucena****

* Swedish Defence Research Agency, Tumba, SE
** Technical University of Vienna, Vienna, AU
*** Technical University of Clausthal, DE
**** University of Malaga, Malaga, ES

sara.wallin@foi.se

Abstract:

A review of standoff detection technologies for explosives has been made. The review is limited to trace detection methods (methods aiming to detect traces from handling explosives or the vapours surrounding a charge due to the vapour pressure of the explosive) rather than bulk detection methods (methods aiming to detect the bulk explosive charge). The requirements for standoff detection technologies are discussed. The technologies discussed are all laser based technologies, such as Laser-Induced Breakdown Spectroscopy (LIBS), Raman, Laser-Induced Fluorescence (LIF) and IR spectroscopies. The review includes novel techniques, not yet tested in realistic environments as well as more mature technologies which have been tested outdoor in realistic environments.

Keywords: *Standoff detection, explosives, review, trace detection*

1. INTRODUCTION

Searching for standoff detection methods for explosives may lead the uninitiated reader to believe that standoff detection techniques for other purposes are readily available for adaption to use for the detection of explosives. For example, there are IR instruments for detection of chemical warfare agents, and LIDAR (Light Detection and Ranging) systems for environmental monitoring [1]. However, when transitioning to detection of explosives, the problem gets more complex than that.

First of all, explosives pose a particularly difficult detection problem because of the low levels of traces available for detection with trace detection methods. The detection technologies need to be very sensitive in order to detect explosives traces. As an example, typical concentrations of NO_2 (a typical chemical for which LIDAR is used in environmental monitoring) in ambient air are in the ppb-range [2]. The vapour pressure of Sulphur Mustard is 14.6 Pa @ 25 °C [3] while the vapour pressure of TNT (an explosive in the mid vapour pressure range) is $4.2 \cdot 10^{-4}$ Pa @ 25 °C [4].

Secondly, achieving the selectivity necessary to uniquely identify an explosive in a background of interferents is difficult. Many chemical agents have atomic compositions including eg. S, P, F and Cl in addition to the common C, H, N and O atoms present in organic molecules. It is also more difficult to obtain well defined spectral features from relatively large molecules like explosives compared to small molecules like NO, NO_2, SO_2, etc. which are often of interest in environmental monitoring.

Detection of explosives is an area with very high needs for research and development. Current capabilities are very limited, both regarding scenarios and technical capability. A technology inventory for explosives detection [5] combined with a view on current and emerging threats has given a view on necessary research and development of explosives detection knowledge and technologies. It is important to keep in mind that different threats require different technical solutions and that a solution that is appropriate for one environment and type of situation is not necessarily good for another situation. Standoff detection poses requirements very different from a check-point type of scenario such as an airport check-point. However, certain development criteria can be specified as universally desirable, such as lower false alarm rate, higher selectivity and sensitivity and with an increased range of detectable threat substances.

In the report *"Existing and Potential Standoff Explosives Detection Techniques"*[6] the authors defined standoff detection of explosives in the following way: *"Standoff explosive detection involves passive and active methods for sensing the presence of explosive devices when vital assets and those individuals monitoring, operating, and responding to the means of detection are physically separated from the explosive device. The physical separation should put the individuals and vital assets outside the zone of severe damage from a potential detonation of the device."* The zone of severe damage varies with scenario and bomb type but they chose 10 m for a pedestrian suicide bomber and 100 m for a vehicle based bomb.

Standoff detection is one of the most wanted capabilities while it is also one of the largest technical challenges. The large distances required pose several physical difficulties: The intensity of the return light decreases inversely with the distance squared, absorption losses in air (wavelength dependent) and scattering losses in air (wavelength dependent).

The technologies with best potential for being fast, selective, sensitive, able to detect many substances and upgradeable to new threats are laser based spectrometric methods for trace detection. Even though some bulk detection methods also have potential for standoff detection they are generally of imaging types that give less information. They may however make an excellent complement to the more specific trace detection technologies by providing information on the presence of electronic devices (maybe part of an Improvised Explosive Device (IED)) or suspicious, concealed objects under clothing. This type of imaging methods is not part of this review.

Although detection limits and the vapour concentrations and particulate amounts that can be expected from explosives are not the focus of this paper, a review of standoff detection methods is incomplete without some discussion about it. Many of the standoff detection methods discussed in this paper have detection limits too high for detection of explosives under realistic conditions outside the laboratory. However, this does not mean that these methods should necessarily be discarded but rather that they should be improved to achieve lower detection limits.

For vapour phase measurements a maximum vapour concentration can be found by examining the vapour pressure of these explosives. However, these concentrations are never achieved in most detection scenarios because equilibrium is not reached. Some explosives, like TATP, have very high vapour pressures (4.3 Pa @ 25°C) [4] and can be expected to be found in high concentrations detectable also with methods with high detection limits.

Other explosives, like HMX, have very low vapour pressure ($5.9 \cdot 10^{-16}$ Pa @ 25°C) [7] and cannot be expected to be detectable with any detection methods in the gas phase. Detection of such explosives must be targeted by detection of particles, also at standoff distances. A critical review of vapour pressures of explosives is to be published soon [8].

2. LASER-INDUCED BREAKDOWN SPECTROSCOPY (LIBS)

Laser Induced Breakdown Spectroscopy (LIBS) is a detection method that uses a laser with high enough energy to break down the sample into plasma. This plasma emits light with characteristic frequencies from ionic, atomic and molecular species that can be detected with a spectrometer, allowing identification of the elemental composition.

DeLucia et al.[9] studied the LIBS spectra of black powder, RDX and air. They concluded that nitrogen and oxygen from the air influenced the spectra of RDX and black powder. They suggested that the ratio between spectral peak intensities could be used for identification. Furthermore, they noted that the laser-induced plasma had never initiated the explosives in their study but that they expected energetic materials with higher sensitivity to shock to ignite.

Noll et al.[10] measured LIBS spectra from ammonium nitrate, DMNB and TNT. They used a tertiary diagram to demonstrate the separation of these materials from each other. However, they do not mention the selectivity in terms of resolution and other molecules with the same atomic composition. A quick search in the NIST Chemistry Webbook [11] reveals that at least for DNMB and TATP, other molecules with the same atomic composition exist. Apart from this obvious source of interferences, there are other components of explosive compositions, surface materials and ambient air that will mix with the LIBS signal from the explosive material itself.

An important question to answer to assess the usefulness of LIBS for explosives detection is whether it is possible to accurately identify the detected species in a real environment. In a real environment there will be many interfering substances. The explosive surface may not be exposed so the detection may be made on trace particles on a surface. Therefore other dust particles and dirt as well as parts of the surface may be interfering with the result. There can also be contributions from nitrogen and oxygen in the air.

To increase the selectivity of LIBS, there are several tricks to be employed. The first is the use of double pulses. The first laser pulse will produce a so called laser generated vacuum [12] allowing the second pulse a few microseconds later to get a cleaner result without the influence of ambient air. This method was employed for explosives by DeLucia et al. [13].They compared data from single pulse LIBS under an Argon atmosphere with data from double pulse LIBS under ambient air. Double pulse operation was made with half the pulse energy and the pulses were separated by 1-10 µs. They measured LIBS spectra from RDX as well as residues of RDX, Comp-B and diesel fuel on aluminium and investigated the O/N ratio as well as the O/C ratio under various conditions. For RDX residues, they found that O/N increased from 2.2±0.2 in air to 13±5 and O/C decreased from 28±9 to 3.7±0.6 when the measurement was repeated under Ar atmosphere. While investigating bulk RDX and switching to double pulse LIBS, O/C decreased from 14±3 for single pulse measurements to 4.9±0.9 when double pulses were employed. Their conclusion from the overall analysis of the results was that interference from atmospheric oxygen and nitrogen will be minimized but not eliminated when employing double pulse LIBS. They demonstrated enhanced selectivity with double pulse LIBS as compared to single pulse LIBS by a simple comparison for the discrimination of RDX and diesel fuel.

Gottfried et al. (same research group as De Lucia et al.) also published a paper [14] where double pulse LIBS was used to detect explosives at a standoff distance of 20 m. They mention that double pulse LIBS also increases the sensitivity by increasing the ablation rate, plasma volume, temperature and ion density, thus enabling larger standoff distance when using double pulse LIBS in addition to the enhanced selectivity. The distance of 20 m was limited by laboratory space rather than sensitivity and they expect that longer standoff distances could be used. Their experimental setup also included a digital camera and a wireless range finder for ease of finding the target as well as measuring the distance to it. They measured the spectra of RDX and Comp-B as well as several other materials. They use Principal Components Analysis (PCA) in order to separate the samples of aluminium, RDX residue, WD-40 oil, Arizona dust and fingerprint residue in a two-dimensional plot. Six ratios of the C, H, N and O peak intensities – H/C, N/C, O/C, N/H, O/H and O/N were calculated for each spectrum, forming the basis for the PCA analysis.

A standoff LIBS system was tested in a field test at at Yuma Proving Ground at distances up to 45 m using a coaxial setup and a single-pulse laser [15]. Peak ratio analysis combined with analysis of the C_2 Swan bands were used for identification. Residues of solid materials were placed at the target (the door of a vehicle) 30 meters away, either by placing a fingerprint or via a solution of the solid in acetone with subsequent evaporation of the solvent. Influences from oxygen and nitrogen in air were observed. Analysis of peak ratios in combination with analysis of molecular bands was used for identification. First, 15 known samples were placed on the door. Then also 6 unknown samples were measured in a blind test. The laser beam was defocused in order to increase the target surface and avoid damage to the door surface. Preliminary data processing was carried out by first distinguishing organic compounds from inorganic compounds by the presence of the C_2 Swan bands as well as an H and an O line. Explosives were then differentiated from other organics by additional nitrogen lines as well as changed intensity ratios of the C_2 bands compared to the O, N and H emission. One single positive result out of 12 shots was regarded enough for a positive response. The results of this preliminary data processing are not explicitly given in the paper. However, the test with known samples prompted the development of a more thorough data processing algorithm. In this algorithm, Na and K lines were included to differentiate explosives from non-explosives. It was stated that this was based on empirical experience from these trials. By using this algorithm all 6 unknown samples in the blind test were correctly identified. However, it is counter-intuitive to include these lines for correct identification of explosives since they are not specific to explosives but rather present as residues from synthesis or handling after synthesis. It could also increase the sensitivity to environmental interferences.

In order to aid identification of explosives by LIBS, Babushok et al. [16] made a kinetic modelling study of the laser-induced plasma of RDX. They pointed out that although the laser energy is released in the sample on a femto- to nanosecond time scale, an enduring plasma is formed lasting for 10-100 µs. The long existence of this plasma is due to chemical reactions and recombination processes sustaining the plasma. The model included initial decomposition of RDX, high/temperature reactions of these decomposition products with the buffer gas (air and argon) as well as reactions of excited states species with decomposition products and buffer gas. The conclusions of this kinetic modelling were that the reaction mechanism leading to the LIBS reaction products on a µs scale can be divided into two stages. During less than the first µs the high electron concentration sustains high populations of the excited states of H, N and O atoms. During the second stage, the temperature decreases substantially. Their analysis of this two/stage behaviour showed that the initial decomposition mechanisms for RDX is only important during the first stage and that the memory of the initial compounds is lost during the second stage, thus leading to the conclusion that compounds can be identified using their unique ratios of the atoms they consist of. This would also suggest that two different compounds with the same stoichiometry would also give the same LIBS response.

De Lucia, Gottfried et al. [17, 18] describe the development of algorithms for classification of substances by LIBS. Gottfried [18] compared many different algorithms. They started with an algorithm similar to that employed by López-Moreno et al. [15]. However, rather than comparing peak ratios to fixed numbers like López-Moreno, they compared peak ratios to peak ratios to discriminate explosives from non-explosives. The algorithm worked relatively well for pure samples. However, they concluded that for real-world applicability with sample mixtures, a more sophisticated algorithm is required.

A linear correlation method was also used. Library spectra were created for RDX, TNT Al, dust and oil. Each sample was then tested against the library with the sample being tested temporarily removed from the library. The sample was then identified as the compound with which it had the highest correlation. Several variations of building the library were attempted but the conclusion was that linear correlation is unlikely to be an effective method for identification of explosives residues on different substrates.

They also discussed the PCA method from which preliminary results were given in a previous paper [14]. The PCA model is built so that different samples are grouped with other samples of the same type. For good discrimination of samples as many as 13 principal components were needed leading to difficulties visualising the results in two or three dimensions. Using the Soft Independent Method of Class Analogy (SIMCA) which consists of a collection of PCA models instead, visualization is easier but many samples were misclassified.

Furthermore, they described results of Partial Least Squares Discriminant Analysis (PLS-DA). PLS-DA was also described by DeLucia (from the same research group)[17]. This paper describes that a large variety of explosives samples, non-explosives samples and mixtures of both explosives and non-explosives were investigated and the PLS-DA model applied to the discrimination of the samples as explosive or non-explosive. Since LIBS has a large shot-to-shot variability, PLS-DA was expected to perform better than PCA. They concluded that PLS-DA has promise for classification of target materials with LIBS but that there are issues that must be considered before it can be implemented. They write that they are currently collecting more data to be included in the model for better discrimination.

The LIBS experiments described above have exclusively used a laser wavelength of 1064 nm for laser-induced breakdown. Waterbury et al. [19] used 266 nm for laser-induced breakdown instead. The breakdown wavelength of 266 nm was chosen to minimise the eye hazard associated with 1064 nm radiation. The nature of LIBS, which is designed to create plasma, thus damaging any surface it investigates to some extent, can never be eye safe when direct hits are concerned. However, using 266 nm, unintentional stray light would be less likely to damage the eye since 266 nm has 600 times higher MPE (Maximum Permissible Exposure) than 1064 nm radiation. Another reason for the choice of 266 nm is the authors' intention to build in future Raman capability to the system. A consequence of the 266 nm choice is also that the 80 mJ laser is not enough to create good plasma by itself. Therefore a CO_2 laser operating at 10.6 µm was focused on the same spot to enhance the LIBS signal from the plasma. The CO_2 laser is delayed by 0.5 µs from the 266 nm pulse. This technique is called Townsend Effect Plasma Spectroscopy (TEPS) and it provides a more than 100 times signal enhancement. This way, the LIBS signal is large enough also at 40 m distance.

In addition to the experimental trick of using double pulses, the selectivity of LIBS can also be enhanced by adding temporal resolution to the LIBS emission analysis. This information can be used to discriminate between e.g. explosives and plastic materials. This has been briefly described by Bauer at al[20].

Baudelet et al. [21] studied time-resolved LIBS from a pure copper surface, nylon and an aspirin tablet in order to investigate the time dependence of the important molecular band from CN and the important atomic lines from O and C. They found that when measuring the LIBS spectrum from the pure copper surface the emission from oxygen starts at a very low level and reaches a maximum after about 100 ns after which the signal shows an exponential decay. This behaviour is consistent with influences to the surrounding ambient air by the LIBS plasma spark. For nylon on the other hand, which contains much oxygen in itself, the signal starts off high from the beginning and immediately starts an exponential decay. This is consistent with the origin of this oxygen signal being the nylon itself rather than interaction with air. Aspirin shows a more complex behaviour with a combination of prompt and delayed oxygen emission.

CN can be formed in the plasma from nitrogen-containing organic molecules. However, it can also be formed by a reaction between molecular C_2 from the plasma spark and react with N_2 from the air forming CN. When studying the time evolution of the CN molecular band at different laser fluences, Baudelet et al. [21] also found that low fluences give high initial CN band intensities, indicative of the formation of CN directly from the organic material under investigation. Higher fluences give more delayed CN intensity, indicating more CN being formed by reaction with nitrogen molecules from the surrounding air.

3. LIDAR (LIGHT DETECTION AND RANGING)

LIDAR measures scattered light to determine the range to a distant target by emitting a short laser pulse and recording the echo (reflections) from objects in the light path. The position of the object is given by the time delay of the echo. The echo from each laser pulse is recorded at multiple times, thus giving information along the beam path.

This basic form of LIDAR does not give any information about the object that is scattering the light. There are however different variations to the LIDAR principle that can give more information. By combining the basic ranging scheme with spectrometric information about the target the chemical composition of the object can be determined. Basically, all pulsed laser based spectrometric methods are suitable for LIDAR. Examples are Raman spectroscopy, LIF and Differential Absorption Lidar (DIAL).

4. RAMAN SPECTROSCOPY

One of the detection methods that are anticipated to meet the requirements for explosives standoff detection is Raman spectroscopy. Raman spectroscopy serves as an analytical tool that gives detailed molecular specific information about the studied molecule. It has been used as a standard analytical tool for identification of chemical substances for many years; however significant advances in laser technology and CCD cameras recently opened the possibility to use this technique also in field measurements. Given its analytical properties based on high molecular selectivity, the building of a large and expandable (to include new threats) reference database containing threat and innocuous substances should be quite feasible.

As opposed to LIF, Raman Spectroscopy is an instantaneous process; an inelastic scattering of photons where some energy is lost to (or gained from) the target molecule, returning scattered light with a different wavelength, the difference corresponding to the energy of vibrational modes of the molecule. These vibrational modes can be regarded as a fingerprint that uniquely identifies the chemical substance or substances in a sample. Also complex mixtures can often be analyzed using algorithms for pattern recognition. A drawback with Raman spectroscopy is the weakness of signal intensity – Raman scattering occurs for one in approximately 10^7 photons incident upon a sample.

The principle of stand-off detection using Raman spectroscopy was first proposed in the mid-1960s [22]. Although theoretically possible the field suffered from controversy regarding the adequacy of sensitivity arising from the low signal levels, long distance ranges and low sample concentrations inherent in stand-off Raman spectroscopy [22]. While early experiments in the 1970s were limited to short stand-off distances and night-time operation a number of different design features were found to be of critical importance in obtaining better signal-to-noise ratios. These early design improvements centred on optimizing the laser and the optical pathway, as well as the careful selection of detection equipment and gating apparatus [22]. These are all features of the spectrometers that have been continually developed since these early experiments to the systems currently in use today that can operate over considerable distances, in daylight and various other environmental conditions.

Today, stand-off Raman detection is used for a wide range of studies where access to the sample can be difficult, dangerous or destructive. Such applications include atmospheric and geological measurements (e.g Sharma [23]) where direct access to the sample is often limited or prohibitively difficult, identification or monitoring of explosives or chemical spills (e.g. Carter [24], Angel[25]) where direct access to the sample could be dangerous or toxic, and studies of ancient artefacts or artworks (e.g. Vandenabeele [26]) where more direct measurements may damage a priceless object.

Prior to 1990, a number of remote Raman systems had been developed for atmospheric measurements of up to a few kilometres in distance but these instruments required high powered lasers and large telescopes for light collection, sometimes up to 36 inches in diameter. Therefore, the size of the components required meant that these systems were either stationary or required large vehicles for transportation [25]. In 1992 Angel et al. [25] published details of their smaller Raman system which was able to measure a range of solid and liquid samples, analogous to substances found in waste tanks, at distances of 6.3 and 16.7 m and collection times of 60 to 120 seconds with both 488 and 809 nm laser excitation [25].

The weak signal intensity of Raman spectroscopy is a complication to its use for trace detection; it makes it sensitive to ambient light and fluorescence from the sample itself or other chemicals in the vicinity. To improve on this, a pulsed laser can be combined with a gated ICCD detector. Carter et al. [27] showed that the increase in gain provided by the ICCD detector also gives rise to a higher noise level, generating similar signal-to-noise ratios as those obtained using conventional CCD cameras. However, when coupled to a pulsed laser system the ability to gate the detector, and thereby exclude a large proportion of the ambient light, reduces the Raman background. Carter et al[24] reports 2 µs as optimal gating time for their system – shorter gating time is adverse due to the electronic jitter of their system, although this could easily be improved. An essential improvement in S/N-ratio from the use of gated detection may appear in situations where samples contain long-lived luminescent or fluorescent components, or in high ambient light conditions[23, 27].

Many portable Raman standoff systems utilizes a pulsed, frequency doubled Nd:YAG-laser of 532 nm. Examples hereof are standoff Raman spectroscopic systems developed for characterization of planetary surfaces [23, 28, 29, 30, 31] at distances of about 5-65 m. Also utilizing 532 nm Nd:YAG laser, Carter et al. [24, 27] have studied the identification of 4-8% solid explosives in dry sand at up to 50 meters. Yet another example is the portable remote Raman system for monitoring environmental pollution and gases on planetary surfaces up to 100 meters by Sharma et al. [32, 33] - they also measured stand-off spectra of explosives at 10 m distance as well as various liquid explosives and precursors at distances up to 100m. A Nd:YAG-laser based instrumentation of 532 nm was also used by Pettersson et al.[34] for the detection of a number of explosives and precursors at 30 meters distance in an outdoors environment. Samples studied included TATP, HMTD, HP, MEKP, NM, NB and IPN. Different container materials were also studied (brown and green glass, PET bottle). Measurements were conducted in full sunshine as well as during heavy snow fall, demonstrating the applicability of 532 nm for gated Raman spectroscopy under harsh weather conditions. In the study it was concluded that indeed the 532 nm wavelength, falling within the visible spectral region, is within a regime that is prone to fluorescence interference. Despite this, the possibility to conduct Raman interrogation with this wavelength could be of advantage because of the insensitivity to absorption by rain/snow in the measurement path and because of its ability to transmit through numerous container materials.

As mentioned, 532 nm is not the optimal Raman probing wavelength since fluorescence is often occurring in this spectral region. In addition, The Raman signal intensity scales with $1/\lambda^4$, leading to the conclusion that UV wavelengths would be the better choice. However, 532 nm falls within the visible spectral region where optical components are easily available and affordable, making it uncomplicated to design the experimental setup or instrumentation. With the use of pulsed lasers and gated detection, the influence from ambient background is neglible even during day time measurements.

The integration of Raman spectroscopy with LIBS (laser induced breakdown spectroscopy) has been considered by Wiens et al. [35] for the purpose of mineralogical analysis. The combination of the techniques are found to be complementary, with LIBS providing information on trace elements and cations in ionic bonds while Raman provides information on anionic species but can also differentiate between minerals with identical chemical composition but separate structures such as calcite and argonite [35]. These qualities have lead to further interest in combination instruments, as can be exemplified by the work of Courreges-Lacoste et al.[36] They consider the synergies of an integrated instrument and present an integrated system concept. Main interest for this instrument is low mass, size and power consumption since the purpose is to take the instrument on the next ESA mission to Mars for organic, mineralogical and elemental characterisation of Martian soil and rock.

While many of the applications of stand-off Raman spectrometers have not been focussed on the detection of explosives, some studies have assessed the applicability of such systems for remote identification of high explosives and associated compounds. Carter et al. [24] have developed a stand-off Raman system capable of detecting explosive compounds at 27 and 50 m. This system utilises a frequency-doubled Nd:YAG laser, producing 532 nm excitation, and a 1μs-10 ms gate width to allow good spectra collection in ambient light conditions. Explosives investigated were RDX, TNT and PETN as well as nitrate- and chlorate-simulants. All explosives were of a concentration of 8% in a silica matrix except the PETN which was of 4% concentration in a silica matrix. Their estimates suggest that such a system is capable of detecting explosives such as RDX in concentrations of 250 parts-per-million or higher when using an integration time of 100 seconds. Carter et al. also studied the effects of the laser power on their explosive samples and found that, for the majority of those studied, there appeared to be no significant laser-induced degradation of the sample and that signal intensities increased linearly with laser power density up to 3×10^6 W/cm^2. However, for TNT, spectroscopic and visual analysis indicated that at laser powers greater than 3.4×10^5 W/cm^2 can induce photo- or thermal degradation of the sample, highlighting the importance of experimental parameters for measuring unknown explosive samples.

5. UV-RAMAN AND RESONANT RAMAN SPECTROSCOPY (RERS)

Several groups have pursued Raman spectroscopy at UV wavelengths. The reasons for doing so are multiple; the obvious one is that Raman signal intensity scales as $1/\lambda^4$, thus leading to an improvement of the inherently weak Raman response upon interrogation.

Another advantage is the much reduced interference of fluorescence. At wavelengths shorter than 250 nm, the full Raman range, up to 4000 cm^{-1}, is normally free from luminescence. This can be used for identifying species in a fluorescent matrix [37]. The problem of ambient light background that can be present for longer gating times or continuous measurements disappears if the measurements are done in the solar blind region, below 300 nm, as the Hartley absorption band of ozone will block solar radiation.

By using a laser wavelength that falls outside the visible spectral region, the problem of eye hazards is significantly decreased since the MPE is markedly higher for wavelengths that are not focused by the lens of the human eye onto the retina. For the spectral region of 180 to 400 nm, American standard allows up to 3 mJ/cm^2.

Above all these advantages of turning to UV wavelengths, perhaps the most intriguing is the possible resonance enhancement that can occur when the probe laser wavelength is nearing an electronic transition within the molecule; this can give rise to enhancements of certain Raman lines (those coupled to the electronic transition) for which the Raman cross section can be significantly enhanced by a factor of $100\text{-}10^6$ (Szymanzki [38] as referenced in Eckbreth [39]) by resonance effects.

Brookhaven National Laboratory has developed a Mobile Raman Lidar Van (MRLV)[40] for identification of bulk chemical spill (surface contaminations, ~500 g/m^2) at distances of 0.5 km or more using quadrupled Nd:YAG lasers at 266 nm and up to 60 s integration time. They have also performed point UV Raman measurements on some explosives[41] in 0,5-1 ‰ concentration using 248 nm laser radiation. At this wavelength they noted a significant near resonance enhancement (*vide infra*) of some of the vibrational modes. In a joint effort with ITT industries, they did further work on a mobile prototype, LISA (Laser Interrogation of Surface Agents), with enhanced performance, aiming at producing a system adequate for military reconnaissance. The performance requirement of such a system is 0.5 g/m^2 of chemical spills. LISA is a LIDAR system designed for distances of less than 50 m. LISA uses a line narrowed excimer laser of 248 nm, which results in much reduced fluorescence as compared to the 266 nm used in MRLV [42].

Several papers discuss near resonance Raman spectroscopy on narcotics, explosives and chemical agents. Lacey and Sands et al. [43, 44] studied explosives and drugs with UV near resonance Raman spectroscopy using a frequency doubled CW Ar ion laser. They compare the Raman spectra of thin films of sample on a microscope slide (exposing typically 1 pg to the laser) when using 633 nm with Raman spectra with UV excitation at 244 nm. They conclude that fluorescence interference which can pose severe problems and even make measurements impossible at 633 nm is eliminated using 244 nm. They also observe resonance enhancement of the Raman bands associated with vibrations involving atoms on which the electrons involved in the electronic excitation are located. In some cases the spectra of similar molecules such as 3,4-DNT and 2,4,6-TNT are very similar due to the simplification of the spectra that the enhancement of only some Raman bands gives. However, they also conclude that the molecules can still be unambiguously identified with good enough resolution (in the order of a cm^{-1}). Nagli and Gaft[45] and Christesen et al.[46] have also studied near resonance Raman on explosives and chemical agents.

Hochenbleicher et al.[47] made a laboratory study for a resonance Raman LIDAR system on various gases. They found that absorption of the Raman scattered light by the scattering gas can reduce the observed resonance Raman signal to normal Raman scattering levels. From their study they concluded that the sensitivity of a Raman LIDAR system can only be significantly enhanced by resonance under three specific conditions:

- A laser with a minimum of 10 mJ at a wavelength coinciding with a strong absorption band of the sought molecule must be available
- The absorption coefficient of this molecule is of the order of 10^3 l·mol^{-1}·cm^{-1} or more.
- The gas is concentrated in a cloud or plume at the emission source and not yet dispersed in a greater distance from the source.

Phillips and Myers[48] studied resonance Raman effects of nitromethane in 5-18 mM solutions in cyclohexane, acetonitrile and water as well as in vapour phase (10 Torr). They observed high overtones in the spectra up to about 15000cm^{-1}.

Some authors have especially commented the propensity of samples to degrade by the applied UV radiation. Ways of countering this phenomenon is to keep pulse energy at a minimum and to continuously move/rotate the sample [43, 49].

Gaft and Nagli [50] have conducted several studies on the application of UV excitation wavelengths for the detection of explosives. They found that, for 532 nm excitation, Raman detection of explosives could be hampered by the presence of other materials in the background such as metal surfaces, plastics or cloth that the explosives may be contained in or deposited on, with such materials giving rise to luminescence [37]. When moving to 355 and 266 nm excitation using conventional laboratory based systems, the third and fourth harmonics of a

Nd:YAG laser, the problem of luminescence from other materials can be compensated for by the increase in Raman cross section of the explosive molecules at UV wavelengths [37]. For a number of explosive materials, with fundamental absorption bands in the UV, the move to UV excitation wavelengths increases the likelihood of generation of resonance or pre-resonance Raman signals which will provide an additional increase in signal intensity [50]. Gaft and Nagli found that for the majority of explosives the signal to noise ratio was much better when using 266 nm than when using 355 nm. The reason for this is likely to be that background luminescence is weaker at shorter wavelengths and situated further away from the Raman signals. However, RDX and its derivatives were the exception to this with a much clearer spectrum recorded using 355 nm, whereas the 266 nm spectrum was significantly poorer. At an even shorter wavelength, 248 nm, RDX spectral features are again clearly visible. The authors also conclude that for severe luminescence from the sample itself, as is the case for the RDX composition Semtex, 248 nm is the only wavelength (out of 532, 355, 266 and 248 nm) that sufficiently discriminates fluorescence to allow identification [37, 50].

A further study [50] has investigated the Raman cross sections of some explosives at different wavelengths in more detail. They conclude significant pre-resonance enhancement of certain Raman lines in the UV spectral range for all explosives studied; UN, TATP, RDX, TNT, and PETN. For the disappearance of RDX spectral features at 355 nm (see above), the authors suggest that the reason could be that numerous electron transitions contribute to polarizability of the molecule; this can result in different resonant electronic transitions changing their intensity with wavelength. For RDX, some typical Raman lines (present at 355 nm) disappear and instead a weak double band appears (for 248 nm) at another spectral position [45].

6. LASERINDUCED FLUORESCENCE (LIF)

Laser-Induced Fluorescence is a commonly used analytical tool for detection and identification of many substances. It is a very valuable tool in combustion diagnostics [51] and for the study of decomposition of explosives [52]. However, when very specific information is required, it is best suited for the analysis of small molecules such as diatomics and triatomics. The larger the molecule gets, the less distinguishing structures are found in the LIF signal. An example can be found in Milton et al. [53] who made DIAL (Differential Absorption LIDAR) measurements on Toluene and Benzene. This is probably a reason why the authors of this paper have been unable to find any LIF work done on the detection of explosives, despite the excellent standoff potential of the method.

7. PHOTO-FRAGMENTATION LASER INDUCED FLUORESCENCE (PF-LIF)

Photo fragmentation (PF) followed by fragment detection is an indirect technique frequently applied for the detection of substances which are difficult to detect spectroscopically with direct techniques. Larger molecules, such as most explosives, often have weak and poorly defined transitions available for spectroscopic detection whereas smaller molecules, such as di- and triatomics, mostly have strong and well defined transitions.

Photo-fragmentation LIF is a variation of LIF exploiting an initial fragmentation of the explosive molecule followed by LIF detection of characteristic fragments. Many explosives have a molecular structure containing nitro-groups and PF-LIF is restricted to use for detection of these explosives. False alarms may occur from innocuous nitro-compounds.

The detection scheme for PF-LIF can be made with slight variations. When exposed to UV laser radiation, NO_2-containing compounds are photo fragmented forming NO_2 which is rapidly predissociated to NO and O. This fragmentation is wavelength dependent [54] but fragmentation occurs in a wide spectral range. Being a diatomic molecule, NO has a characteristic spectrum in the UV and visible spectral regions. LIF is used to detect NO.

Lemire et al. [55] performed experiments based on photofragmentation primarily followed by Resonant Enhanced Multiphoton Ionization Mass Spectrometry (REMPI-MS) but also followed by LIF for comparison for one substance, DMNA (dimethylnitramine). The LIF detection of NO went through excitation in the (0,0) transition of $A^2\Sigma^+ \leftarrow X^2\Pi$. They obtained a 2 ppm detection limit with PF-LIF, which was 4 times less sensitive than PF-REMPI-MS. They concluded that high backgrounds in the LIF measurement may have caused the difference in sensitivity.

Swayambunathan et al [56] made similar experiments on PETN, RDX and TNT, also comparing PF-REMPI and PF-LIF using the same transition. They achieved a 37 ppm detection limit with PF-LIF as compared to 70 ppb for PF-REMPI. They conclude that this difference is due to limitations in their system design as well as higher impact of collisional quenching of NO in the excited state by N_2 and O_2 in the PF-LIF experiment.

Wu et al. [57] studied PF-LIF on TNT samples, pure and in simulated soil samples, and its dependence on pressure, temperature and temperature cycling. They found that repeated heating of the TNT sample reduced the PF-LIF signal intensity with the number of heating cycles. They also found that the optimal pressure giving the highest PF-LIF response was about 400 Torr, i.e. well below ambient pressure. They estimated that the limit of detection of TNT in soil is in the order of 40 ppb.

Arusi-Parpar et al. [58] used another PF-LIF detection scheme. By exciting NO in the (0,2) transition of $A^2\Sigma^+ \leftarrow X^2\Pi$ rather than the (0,0) transition, they avoid background signals from vibrationally relaxed NO in the ground state naturally present in ambient air. The amount of NO naturally present in the air is in the order of ppb levels, i.e. high enough to disturb the detection of NO fragments from explosives in ambient conditions. They achieved a detection limit of less than 8 ppb for TNT. As pointed out by Cabalo et al [59], the detection of vibrationally excited NO is dependent on detection being made in a very short time (~10 ns) after the fragmentation as NO will otherwise have time to vibrationally relax. Since the atmospheric NO is not excited above $\upsilon'' = 0$ to any noticeable extent at room temperature, atmospheric NO interference is significantly reduced (Boltzmann distribution gives the relative populations of NO in the $\upsilon'' = 0, 1, 2$ states as $1, 10^{-4}, 10^{-8}$, respectively, at room temperature). In dissociated TNT molecules, at least 30% NO is produced with a $\upsilon'' = 0, 1, 2$ ratio of 1, 0.5, 0.1, respectively. Thus, in this detection scheme, interference from atmospheric NO is greatly reduced. Another contributing factor for interference reduction is that the fluorescence is measured at shorter wavelengths than the excitation wavelength, rejecting fluorescence from TNT and other molecules in the air.

A year later, Parpar et al. had adapted the experiment to standoff detection at 2,5 m [60]. They report detection limits in the low ppb region. However, as they point out, they cannot give any absolute value to the concentration in their experiment as TNT flakes are placed on a temperature controlled sample holder in ambient air resulting in significantly lower concentrations near the surface than the temperature would indicate for equilibrium conditions. They determined the response time of their system to 15 seconds.

After introducing a method to remotely evaporate an explosive sample from a surface, their results were improved by two to three orders of magnitude giving much higher S/N while shortening the measurement time significantly (from 100 s per measurement point to 10 s)[61]. By detecting the presence of nitro-explosives by simply measuring on and off the NO LIF signal the measurement time can be shortened to between 2 and 10 seconds. They report that the major reason that it is not shorter is the turbulent nature of the evaporation process leading to strong variations in vapour concentration above the sample. They predicted that standoff distances up to tens of meters should be possible while maintaining the same S/N.

Cabalo et al. [59] detected explosives residues in particulate form by fragmentation of explosives directly from the surface and subsequent detection of the photofragments. They chose to detect the fragments by REMPI rather than LIF and achieved detection limits in the low ng/cm^2 region. By comparing the detection limits achieved for RDX, HMX, CL-20 and TNT respectively they found that the explosives with highest R-NO$_2$ bond dissociation energy also had the highest limit of detection as could be expected.

8. PULSED LASER FRAGMENTATION MID-INFRARED SPECTROSCOPY

Explosives as well as their decomposition products possess strong absorption features in the mid-infrared (MIR) spectral region between $\lambda = 5$ and 11 μm [62, 63, 64]. Due to the strong absorption features, directly related to the molecular structure of molecules, between 3 and 12 μm the MIR spectral region is also known as the "fingerprint region".

However, besides strong absorption features spectroscopic detection relies on the presence of the substance to be investigated in the gas phase. This is an obstruction for the detection of most explosives since their vapour pressures are extremely low. For example, TNT possesses an equilibrium vapour pressure $4.2 \cdot 10^{-4}$ Pa @ 25°C[4] and RDX one of $3.0 \cdot 10^{-11}$ Pa[7]. Therefore, optical methods are only applied so far for the detection of these substances at atmospheric pressure in the MIR region after sampling and pre-concentration. Triacetone triperoxide (TATP), an explosive commonly used for terrorist attacks has a much higher vapour pressure leading to a higher concentration in air at ambient conditions. Oxley et al. [4] obtained a room temperature vapour pressure of 0.4 Pa for TATP.

The detection of nitro-based explosives with lower vapour pressure has been made using pulsed laser fragmentation (PLF) to produce a measurable concentration of decomposition products that are subsequently detected by synchronized pulsed MIR laser absorption spectroscopy[65, 66, 67]. The detection is performed in two steps [65]. First, pulsed laser fragmentation (PLF) is applied and then MIR absorption spectroscopy with a synchronized probe Quantum Cascade Laser (QCL) is used for sensing of constituents of the generated fragments of the surface layer of the sample to be investigated in a second step. Different databases (e.g., HITRAN [68]) provide information on line strength and position of different molecules that can be used for a comparison with experimental results.

During PLF, molecules from the surface layer are fragmented forming a plume very close to the surface. The plume generated by PLF contains NO$_x$ molecules. Repeated laser shots to the surface are used for several minutes to build up enough NO for detection. Willer et al. [67] demonstrated that the NO concentration was continuously built up over time for HMX but remained relatively constant for TNT (which has fewer nitro-groups than HMX).

Bauer et al. [65] used chemiluminescence measurements to determine the NO and NO$_2$ concentrations in the PLF plume after 30 seconds. They concluded that for the materials that they tested (TNT, aluminium, plastics (polyamide) and TNT contaminated polyamide) the ratio between NO and NO$_2$ concentrations can be used to determine if the material is an explosive or a plastic material. They also concluded, using an uncontaminated aluminium surface, that with their experimental setup, no NO$_x$ emissions from ambient air reactions contribute to the plasma.

Furthermore, due to the availability of powerful and compact Nd:YAG lasers, they performed PLF using 1064 nm. However, investigations with explosives show an improved interaction of light and sample by use of wavelengths around 1.5 μm for PLF [20]. This wavelength is in the so called eye safe region which means that low intensity stray light will not hurt innocent bystanders or the operator.

In later work, Bauer et al. [66] revisited the study of NO/NO$_2$ ratios for plastic materials (PMMA) and explosives (TNT), now a using 1.5 μm laser for PLF. They found that they can discriminate between a non-contaminated PMMA sample and a PMMA sample with a surface contamination of TNT.

In the earlier experiment Bauer et al. [65] detected HMX contamination on a metal plate over a distance of 5m. The telescope optics restricted the distance for stand-off detection to this range but they concluded that with modified optics distances up to 20m seem to be a realistic goal. They also pointed out that water absorption is an issue when using MIR absorption at long distances. Of the NO lines in the region of interest, the one at 1896.2 cm^{-1} is most clear from water absorption.

9. DIAL (DIFFERENTIAL ABSORPTION LIDAR)

DIAL is a technique based on the differential absorption of backscattered light. A pulsed laser beam of two different wavelengths is directed on the target. Some of this laser light is backscattered by dust and aerosol particles towards the detector. A wavelength matching the absorption of molecules along the path will be absorbed while a wavelength not matching the absorption will not. A differential return signal will indicate the presence of the target substance.

DIAL has been extensively used for environmental monitoring [69]. The authors of this paper have however not been able to find any references to work on explosives. A reason could be that the spectral region where explosives have the most distinct spectral features is IR and this is a region in which the application of DIAL is difficult [70]. A reason is the large number of air pollutants which have absorption lines in IR but also availability of suitable lasers in this spectral region. The two wavelengths, on and off absorption of the target molecule, must be chosen carefully to avoid interferences from other molecules. Jeffers et al. [71] made diode laser measurements of benzene in vapour phase and demonstrated that it is possible to find a window with a benzene peak free from interferences from ambient air. The detection limit was 1 ppmv.

10. COHERENT ANTI-STOKES RAMAN SPECTROSCOPY (CARS)

Conventional Raman spectroscopy has the disadvantage of the inherent weakness of the incoherent scattering effect. Coherent anti-stokes Raman spectroscopy (CARS) is a variant of Raman spectroscopy typically used in combustion diagnostics because of its good spatial and temporal resolution. The CARS technique is based on a nonlinear conversion of three laser beams into a coherent laser-like Raman beam of high intensity in the anti-Stokes wavelength region.

Peter et al. [72] present a technique of using one broadband laser as two broadband sources. A third narrowband source then defines the spectral resolution. In this way, a spectrum covering the full range of the broad band sources (> 3000 cm^{-1}) can be collected in a single laser shot. The authors report spectral resolution of < 1cm^{-1}, spatial resolution <0.05 mm^2 and temporal resolution < 1s. CARS spectroscopy has been used for combustion diagnostics for many years but has only recently been applied to explosives detection.

Portnov et al. [73] used Coherent anti-Stokes Raman scattering spectroscopy (CARS) for detection of solid particles of eg. RDX. The CARS spectra were recorded in the fingerprint region and were shown to exhibit the strong characteristic features of spontaneous Raman spectra of the respective compounds. Katz et al. [74] has demonstrated stand-off detection of trace amounts of solid explosives (KNO$_3$, RDX) by nonlinear Raman spectroscopy (backscatter

CARS) using shaped femtosecond pulses at ranges up to 12 m. Li et al. [75, 76] used Coherent mode-selective Raman excitation to examine the possibility of standoff detection. They report the detection of characteristic Raman lines for several chemicals using a single-beam coherent anti-Stokes Raman scattering (CARS) technique from a 12 meter standoff distance. Single laser shot spectra were obtained with sufficient signal to noise ratio to allow molecular identification. Background and spectroscopic discrimination were achieved through binary phase pulse shaping for optimal excitation of a single vibrational mode.

11. NOVEL TECHNIQUES

A number of novel techniques for standoff detection of explosives have also been suggested, such as stand-off superradiant spectroscopy (SOS) [77, 78], photo-acoustic spectroscopy [79], differential reflection spectroscopy (DRS) [80, 81, 82] and Resonant infrared photo-thermal imaging [83].

Using mid-IR laser based quartz enhanced photo-acoustic spectroscopy Neste et al [84] recently demonstrated the possibility to identify explosives such as RDX, TNT and PETN at stand-off distances from 2 – 20 m. In their experiments two tuneable, external cavity Quantum Cascade Lasers, pulsed at different rates, were used for illuminating the sample surface. The reflected/scattered light was collected and focused using a mirror and detected acoustically using tuning forks whose frequencies match the pulse rates of the individual lasers. Plots of the amplitudes of the mechanical resonance frequencies of the tuning forks as a function of illuminating wavelength showed good agreement with the traditionally recorded infrared spectra of the analytes. Based on this approach a sensitivity of 100 ng/cm^2 for surface adsorbed analytes such as the studied explosives and tributyl phosphate was reported.

In Resonant infrared photo-thermal imaging an infrared (IR) laser is directed to a surface of interest, which is viewed using a thermal imager. Resonant absorption by the analyte at specific IR wavelengths selectively heats the analyte, providing a thermal contrast with the substrate. Very little results of applying these techniques have been reported. Nevertheless they are interesting and merits further investigation.

12. DISCUSSION

None of the technologies reviewed in this paper could be described as ready for deployment or even for full functionality prototype manufacturing. They all need further R&D work to be useful in a real environment outside the laboratory. However, some of them are more mature than others. Standoff detection is an area of research under rapid development. This can be seen from the vast majority of papers reviewed in this paper being produced after the year 2004. Publications on explosives for some of the newest technologies have only started being made in 2008.

As discussed in the introduction to this paper, many of the standoff detection methods discussed in this paper have detection limits too high for detection of explosives under realistic conditions outside the laboratory. Many also have problems with their selectivity not being good enough to handle the interferents of the environment outside the laboratory.

This review has discussed many standoff detection technologies for explosives. Some of these, like LIBS, Raman and PF-LIF have been tested outside the laboratory and should be considered more mature technologies even though they are still at the research stage. Some of the detection technologies measure molecular properties in the vapour phase, others measure molecular properties directly from particulate form and some of them measure the properties of fragments of these molecules. Therefore, direct comparisons of these technologies have not been made.

Although further R&D can probably increase both sensitivity and selectivity of some of these methods, it is unlikely that one technology will be the sole solution. Integration of several technologies, using their respective strengths is however very likely to be a fruitful way forward.

Acknowledgment:
Part of the research leading to these results has received funding from the European Community's Seventh Framework Programme (FP7/2007-2013) under Grant Agreement n°218037.

13. REFERENCES

[1] P. WEIBRING, H. EDNER, S. SVANBERG: Versatile Mobile Lidar System for Environmental Monitoring, Applied Optics, No.42 p. 3583-3594, 2003.
[2] The Conference Board of Canada: Environment - Urban Nitrogen Dioxide Concentration, http://www.conferenceboard.ca/HCP/Details/Environment/urban-nitrogen-dioxide-concentration.aspx,
[3] L. K. ENGMAN, A. LINDBLAD, A.-K. TUNEMALM, O. CLAESSON, B. LILLIE-HÖÖK, 2002.
[4] J. C. OXLEY, J. L. SMITH, K. SHINDE, J. MORAN: Determination of the Vapor Density of Triacetone Triperoxide (TATP) Using a Gas Chromatography Headspace Technique Propellants Explosives Pyrothechnics, No.30, p. 127-130, 2005.
[5] A. PETTERSSON, S. WALLIN, B. BRANDNER, C. ELDSÄTER, E. HOLMGREN: Explosives Detection – A Technology Inventory, FOI, Swedish Defence Research Agency, FOI-R--2030--SE, Stockholm, 2006
[6] Existing and Potential Standoff Explosives Detection Techniques, Committee on the Review of Existing and Potential Standoff Explosives Detection Techniques, 0-309-09130-6, Washington D.C., 2004
[7] R. B. CUNDALL, T. F. PALMER, C. E. C. WOOD: Vapour Pressure Measurements on Some Organic High Explosive, J. Chem. Soc. Faraday Trans No.74, p. 1339-1345, 1978.
[8] H. ÖSTMARK: To be published, 2009
[9] F. C. DE LUCIA, R. S. HARMON, K. L. MCNESBY, R. J. WINKEL, A. W. MIZIOLEK: Laser-induced breakdown spectroscopy analysis of energetic materials, Applied Optics, No.42, p. 6148-6152, 2003.
[10] R. NOLL, C. FRICKE-BEGEMANN: Stand-Off Detection of Surface Contaminations with Explosives Residues Using Laser-Spectroscopic Methods, (Stand-off Detection of Suicide Bombers and Mobile Subjects, H. Schubert, A. Rimski-Korsakov, Eds.), Springer Netherlands, pp. 89-99, 2006.
[11] NIST, National Institute of Standards and Technology: NIST Chemistry WebBook, http://webbook.nist.gov/chemistry/, 2009-02-15
[12] V. I. BABUSHOK, F. C. DELUCIA, J. L. GOTTFRIED, C. A. MUNSON, A. W. MIZIOLEK: Double pulse laser ablation and plasma: Laser induced breakdown spectroscopy signal enhancement, Spectrochimica Acta Part B-Atomic Spectroscopy, No.61, p. 999-1014, 2006.

[13] F. C. DE LUCIA, J. L. GOTTFRIED, C. A. MUNSON, A. W. MIZIOLEK: Double pulse laser-induced breakdown spectroscopy of explosives: Initial study towards improved discrimination, Spectroschimica Acta Part B, No.62, p. 1399-1404, 2007.
[14] J. L. GOTTFRIED, F. C. DE LUCIA, C. A. MUNSON, A. W. MIZIOLEK: Double-pulse standoff laser-induced breakdown spectroscopy for versatile hazardous materials detection, Spectrochimica Acta Part B, No.62, p. 1405-1411, 2007.
[15] C. LÓPEZ-MORENO, S. PALANCO, J. J. LASERNA, F. C. DELUCIA, A. W. MIZIOLEK, J. ROSE, R. A. WALTERS, A. I. WHITEHOUSE: Test of a stand-off laser-induced breakdown spectroscopy sensor for the detection of explosive residues on solid surfaces, Journal of Analytical Atomic Spectrometry, No.21, p. 55-60, 2006.
[16] V. I. BABUSHOK, F. C. DELUCIA, P. J. DAGDIGIAN, J. L. GOTTFRIED, C. A. MUNSON, M. J. NUSEA, A. W. MIZIOLEK: Kinetic modeling study of the laser-induced plasma plume of cyclotrimethylenetrinitramine (RDX), Spectrochimica Acta Part B, No. 62, p. 1321-1328, 2007.
[17] F. C. DE LUCIA, J. L. GOTTFRIED, C. A. MUNSON, A. W. MIZIOLEK: Multivariate analysis of standoff laser-induced breakdown spectroscopy spectra for classification of explosive-containing residues, Applied Optics, No.47, p. G112-G121, 2008.
[18] J. L. GOTTFRIED, F. C. DE LUCIA, C. A. MUNSON, A. W. MIZIOLEK: Strategies for residue explosives detection using laser-induced breakdown spectroscopy, Journal of Analytical Atomic Spectrometry, No.23, p. 205-216, 2008.
[19] R. D. WATERBURY, A. PAL, D. K. KILLINGER, J. ROSE, E. L. DOTTERY, G. ONTAI, in *Chemical, Biological, Radiological, Nuclear and Explosives (CBRNE) Sensing IX, Vol. 6954* (Eds.: A. W. F. III, P. J. Gardner), pp. 695409-695401 - 695409-695405, 2008.
[20] C. BAUER, J. BURGMEIER, C. BOHLING, W. SCHADE, G. HOLL: Mid-Infrared Lidar for Remote Detection of Explosives, (Stand-off Detection of Suicide Bombers and Mobile Subjects), Springer, Berlin, 2006.
[21] M. BAUDELET, M. BOUERI, J. YU, S. S. MAO, V. PISCITELLI, X. MAO, R. E. RUSSO: Time-resolved ultraviolet laser-induced breakdown spectroscopy for organic material analysis, Spectrochimica Acta Part B, No.62, p. 1329-1334, 2007.
[22] T. HIRSCHFELD, E. R. SCHILDKRAUT, H. TANNENBAUM, D. TANNENBAUM: Remote spectroscopic analysis of ppm-level air pollutants by Raman spectroscopy, Applied Physics Letters, No.22, p. 38-40, 1972.
[23] S. K. SHARMA, P. G. LUCEY, M. GHOSH, H. W. HUBBLE, K. A. HORTON: Stand-off Raman spectroscopic detection of minerals on planetary surfaces, Spectrochimica Acta Part A No.59, p. 2391-2407, 2003.
[24] J. C. CARTER, S. M. ANGEL, M. LAWRENCE-SNYDER, J. SCAFFIDI, R. E. WHIPPLE, J. G. REYNOLDS: Standoff detection of high explosive materials at 50 meters in ambient light conditions using a small Raman instrument, Applied Spectroscopy, No.59, p. 769-775, 2005.
[25] S. M. ANGEL, T. J. KULP, T. M. VESS: REMOTE-RAMAN SPECTROSCOPY AT INTERMEDIATE RANGES USING LOW-POWER CW LASERS, Applied Spectroscopy, No.46, p. 1085-1091, 1992.
[26] P. VANDENABEELE, K. CASTRO, M. HARGREAVES, L. MOENS, J. M. MADARIAGA, H. G. M. EDWARDS: Comparative study of mobile Raman instrumentation for art analysis, Analytica Chimica Acta, No.588, p. 108-116, 2007.
[27] J. C. CARTER, J. SCAFFIDI, S. BURNETT, B. VASSER, S. K. SHARMA, S. M. ANGEL: Stand-off Raman detection using dispersive and tunable filter based systems, Spectrochimica Acta Part A, No.61, p. 2288-2298, 2005.

[28] P. G. LUCEY, T. F. COONEY, S. K. SHARMA: A Remote Raman Analysis System for Planetary Landers, Lunar and Planetary Science, No.XXIX, 1998.
[29] S. K. SHARMA, G. H. BEALL, H. W. HUBBLE, A. K. MISRA, C. H. CHIO, P. G. LUCEY: Telescopic Raman Measurements of Glasses of Mineral Compositions to a Distance of 10 Meters, Lunar and Planetary Science, No.XXXIV, 2003.
[30] S. K. SHARMA, A. WANG, L. A. HASKIN: Remote Raman Measurements of Minerals with Mars Microbeam Raman Spectrometer (MMRS), Lunar and Planetary Science No.XXXVI, 2005.
[31] A. K. MISRA, S. K. SHARMA, P. G. LUCEY: Single Pulse Remote Raman Detection of Minerals and Organics under Illuminated Conditions from 10 Meters Distance, Lunar and Planetary Science No.XXXVI 2005.
[32] S. K. SHARMA, A. K. MISRA, B. SHARMA: Portable remote Raman system for monitoring hydrocarbon, gas hydrates and explosives in the environment, Spectrochimica Acta Part A, No.61, p. 2404-2412, 2005.
[33] S. K. SHARMA, A. K. MISRA, P. G. LUCEY, R. C. F. LENTZ, C. H. CHIO, in *Proc. SPIE, Vol. 6554*, p. 655405, 2007.
[34] A. PETTERSSON, I. JOHANSSON, S. WALLIN, M. NORDBERG, H. ÖSTMARK: Standoff Detection of Explosives in a Realistic Environment, Propellants Explosives Pyrothechnics, No.Accepted for publication, 2009.
[35] R. C. WIENS, S. K. SHARMA, J. THOMPSON, A. MISRA, P. G. LUCEY: Joint analyses by laser-induced breakdown spectroscopy (LIBS) and Raman spectroscopy at stand-off distances, Spectrochimica Acta Part A, No.61, p. 2324-2334 2005.
[36] G. B. COURRÈGES-LACOSTE, B. AHLERS, F. R. PÉREZ: Combined Raman spectrometer/laser-induced breakdown spectrometer for the next ESA mission to Mars, Spectrochimica Acta Part A, No.68, p. 1023-1028, 2007.
[37] M. GAFT, L. NAGI: UV gated Raman spectroscopy for standoff detection of explosives, Optical Materials, No.30, p. 1739-1746, 2008.
[38] N. A. SZYMANSKI: Raman Spectroscopy, Theory and Practice, Plenum, New York, **1967**.
[39] A. C. ECKBRETH: Laser Diagnostics for Combustion Temperature and Species, Abacus Press, Cambridge **1988**.
[40] M. WU, M. RAY, K. H. FUNG, M. W. RUCKMAN, D. HARDER, A. J. SEDLACEK: Stand-off detection of chemicals by UV Raman spectroscopy, Applied Spectroscopy, No.54, p. 800-806, 2000.
[41] A. J. SEDLACEK III, S. CHRISTESEN, T. CHYBA, P. PONSARDIN, in *Proceedings of SPIE, Vol. 5269*, pp. 23-33, 2004.
[42] T. H. CHYBA, E. AL.: Laser Interrogation of Surface Agents (LISA) for Standoff Sensing of Chemical Agents, 21st International Radar Conference (ILRC), Quebec, Canada, 2002
[43] R. J. LACEY, I. P. HAYWARD, H. S. SANDS, D. N. BATCHELDER, in *Proc. of SPIE, Vol. 2937*, pp. 100-105 1997.
[44] H. S. SANDS, I. P. HAYWARD, T. E. KIRKBRIDE, R. BENNETT, R. J. LACEY, D. N. BATCHELDER: UV-Excited Resonance Raman Spectroscopy of Narcotics and Explosives, Journal of Forensic Science, No.43, p. 509-513, 1998.
[45] L. NAGLI, M. GAFT, in *Proc. of SPIE, Vol. 6552*, p. 65520Z 2007.
[46] S. D. CHRISTESEN, J. M. LOCHNER, A. M. HYRE, D. K. EMGE, in *Proc. of SPIE, Vol. 6218*, p. 621809 2006.
[47] J. G. HOCHENBLEICHER, W. KIEFER, J. BRANDMÜLLER: A Laboratory Study for a Resonance Raman Lidar System, Applied Spectroscopy, No.30, p. 528-531, 1976.

[48] D. L. PHILLIPS, A. B. MYERS: High Overtone Resonance Raman Spectra of Photodissociating Nitromethane in Solution, Journal of Physical Chemistry, No.95, p. 7164-7171, 1991.
[49] N. TARCEA, E. AL.: UV Raman spectroscopy - A tecnique for biological and mineralogi-cal in situ planetry studies, Spectrochimica Acta Part A, No.68, p. 1029-1035, 2007.
[50] L. NAGLI, M. GAFT, Y. FLEGER, M. ROSENBLUH, pp. 1747-1754, 2008.
[51] D. R. CROSLEY, G. P. SMITH: Laser-induced fluorescence spectroscopy for combustion diagnostics, Optical Engineering, No.22, p. 545-553, 1983.
[52] H. ÖSTMARK, M. CARLSON, K. EKVALL: Concentration and temperature measurements in a laser-induced high explosive ignition zone .1. LIF spectroscopy measurements, Combustion and Flame, No.105, p. 381-390, 1996.
[53] M. J. T. MILTON, P. T. WOODS, B. W. JOLLIFFE, N. R. W. SWANN, T. J. MCILVEEN: Measurements of Toluene and Other Aromatic Hydrocarbons by Differential-Absorption LIDAR in the Near-Ultraviolet, Applied Physics B, No.55, p. 41-45, 1992.
[54] J. B. SIMEONSSON, R. C. SAUSA: Laser photofragmentation/fragment detection techniques for chemical analysis of the gas phase, Trac-Trends in Analytical Chemistry, No.17, p. 542-550, 1998.
[55] G. W. LEMIRE, J. B. SIMEONSSON, R. C. SAUSA: Monitoring Of Vapor-Phase Nitro-Compounds Using 226-Nm Radiation - Fragmentation With Subsequent No Resonance-Enhanced Multiphoton Ionization Detection, Analytical Chemistry, No.65, p. 529-533, 1993.
[56] V. SWAYAMBUNATHAN, G. SINGH, R. C. SAUSA: Laser photofragmentation-fragment detection and pyrolysis-laser-induced fluorescence studies on energetic materials, Applied Optics, No.38, p. 6447-6454, 1999.
[57] D. D. WU, J. P. SINGH, F. Y. YUEH, D. L. MONTS, pp. 3998-4003, 1996.
[58] T. ARUSI-PARPAR, D. HEFLINGER, R. LAVI: Photodissociation followed by laser-induced fluorescence at atmospheric pressure and 24 degrees C: a unique scheme for remote detection of explosives, Applied Optics, No.40, p. 6677-6681, 2001.
[59] J. B. CABALO, R. C. SAUSA: Explosive Residue Detection by Laser Surface Photofragmentation-Fragment Detection Spectroscopy: II. *In Situ* and Real-time Monitoring of RDX, HMX, CL20, and TNT, by an Improved Ion Probe, Army Research Laboratory, ARL-TR-3478, Aberdeen Proving Ground,MD, 2005
[60] D. HEFLINGER, T. ARUSI-PARPAR, Y. RON, R. LAVI: Application of a unique scheme for remote detection of explosives, Optics Communications, No.204, p. 327-331, 2002.
[61] T. ARUSI-PARPAR, I. LEVY, in *NATO Advanced Research Workshop on Stand-Off Detection of Suicide Bombers and Mobile Subjects* (Eds.: H. Schurbert, A. RimskiKorsakov), Pfinztal, GERMANY, pp. 59-67, 2005.
[62] I. DUNAYEVSKIY, A. TSEKOUN, M. PRASANNA, R. GO, C. K. N. PATEL: High-sensitivity detection of triacetone triperoxide (TATP) and its precursor acetone, Applied Optics, No.46, p. 6397-6404, 2007.
[63] J. JANNI, B. D. GILBERT, R. W. FIELD, J. I. STEINFELD: Infrared absorption of explosive molecule vapors, Spectrochimica Acta Part A: Molecular and Biomolecular Spectroscopy, No.53, p. 1375-1381, 1997.
[64] M. W. TODD, R. A. PROVENCAL, T. G. OWANO, B. A. PALDUS, A. KACHANOV, K. L. VODOPYANOV, M. HUNTER, S. L. COY, J. I. STEINFELD, J. T. ARNOLD: Application of mid-infrared cavity-ringdown spectroscopy to trace explo-

sives vapor detection using a broadly tunable (6–8 μm) optical parametric oscillator Applied Physics B: Lasers and Optics, No.75, p. 367-376 2002.

[65] C. BAUER, P. GEISER, J. BURGMEIER, G. HOLL, W. SCHADE: Pulsed laser surface fragmentation and mid-infrared laser spectroscopy for remote detection of explosives, Applied Physics B-Lasers and Optics, No.85, p. 251-256, 2006.

[66] C. BAUER, A. K. SHARMA, U. WILLER, J. BURGMEIER, B. BRAUNSCHWEIG, W. SCHADE, S. BLASER, L. HVOZDARA, A. MULLER, G. HOLL: Potentials and limits of mid-infrared laser spectroscopy for the detection of explosives, Applied Physics B-Lasers and Optics, No.92, p. 327-333, 2008.

[67] U. WILLER, M. SARAJI, A. KHORSANDI, P. GEISER, W. SCHADE: Near- and mid-infrared laser monitoring of industrial processes, environment and security applications, Optics and Lasers in Engineering, No.44, p. 699-710, 2006.

[68] L. S. ROTHMAN, D. JACQUEMART, A. BARBE, D. C. BENNER, M. BIRK, L. R. BROWN, M. R. CARLEER, C. C. JR, K. CHANCE, L. H. COUDERT, V. DANA, V. M. DEVI, J.-M. FLAUD, R. R. GAMACHE, A. GOLDMAN, J.-M. HARTMANN, K. W. JUCKS, A. G. MAKI, J.-Y. MANDIN, S. T. MASSIE, J. ORPHAL, A. PERRIN, C. P. RINSLAND, M. A. H. SMITH, J. TENNYSON, R. N. TOLCHENOV, R. A. TOTH, J. V. AUWERA, P. VARANASI, G. WAGNER: The HITRAN 2004 molecular spectroscopic database, Journal of Quantitative Spectroscopy &Radiative Transfer No.96 p. 139-204, 2005.

[69] M. W. SIGRIST: Air Monitoring by Spectroscopic Techniques, Wiley-IEEE, **1994**.

[70] S. AMORUSO, A. AMODEO, M. ARMENANTE, A. BOSELLI, L. MONA, M. PANDOLFI, G. PAPPALARDO, R. VELOTTA, N. SPINELLI, X. WANG: Development of a tunable IR lidar system, Optics and Lasers in Engineering, No.37, p. 521-532, 2002.

[71] J. D. JEFFERS, C. B. ROLLER, K. NAMJOU, M. A. EVANS, L. MCSPADDEN, J. GREGO, P. J. MCCANN: Real-Time Diode Laser Measurements of Vapor-Phase Benzene, Analytical Chemistry, No.76, p. 424-432, 2004.

[72] P. C. CHEN, C. C. JOYNER, S. T. PATRICK, K. F. BENTON: High-Speed High-Resolution Gas-Phase Raman Spectroscopy, Analytical Chemistry, No.74, p. 1618-1623, 2002.

[73] A. PORTNOV, S. ROSENWAKS, I. BAR: Detection of particles of explosives via backward coherent anti-Stokes Raman spectroscopy, Applied Physics Letters, No.93, p. 3, 2008.

[74] O. KATZ, A. NATAN, Y. SILBERBERG, S. ROSENWAKS: Standoff detection of trace amounts of solids by nonlinear Raman spectroscopy using shaped femtosecond pulses, Applied Physics Letters, No.92, 2008.

[75] H. W. LI, D. A. HARRIS, B. XU, P. J. WRZESINSKI, V. V. LOZOVOY, M. DANTUS: Coherent mode-selective Raman excitation towards standoff detection, Optics Express, No.16, p. 5499-5504, 2008.

[76] H. W. LI, D. A. HARRIS, B. XU, P. J. WRZESINSKI, V. V. LOZOVOY, M. DANTUS: Standoff and arms-length detection of chemicals with single-beam coherent anti-Stokes Raman scattering, Applied Optics, No.48, p. B17-B22, 2009.

[77] G. O. ARIUNBOLD, M. M. KASH, H. LI, V. A. SAUTENKOV, Y. V. ROSTOVTSEV, G. R. WELCH, M. O. SCULLY, in *38th Winter Colloquium on the Physics of Quantum Electronics*, Snowbird, UT, pp. 3273-3281, 2008.

[78] V. KOCHAROVSKY, S. CAMERON, K. LEHMANN, R. LUCHT, R. MILES, Y. ROSTOVTSEV, W. WARREN, G. R. WELCH, M. O. SCULLY: Gain-swept superradiance applied to the stand-off detection of trace impurities in the atmosphere, Proceed-

ings of the National Academy of Sciences of the United States of America, No.102, p. 7806-7811, 2005.
[79] C. W. VAN NESTE, L. R. SENESAC, T. THUNDAT: Standoff photoacoustic spectroscopy, Applied Physics Letters, No.92, 2008.
[80] A. M. FULLER, R. E. HUMMEL, C. SCHOLLHORN, P. H. HOLLOWAY, in *Conference on Chemical and Biological Sensors for Industrial and Environmental Monitoring II* (Eds.: S. D. Christesen, A. J. Sedlacek, J. B. Gillespie, K. J. Ewing), Boston, MA, pp. U289-U299, 2006.
[81] C. SCHOLLHORN, A. M. FULLER, J. GRATIER, R. E. HUMMEL, in *Conference on Chemical and Biological Sensing VIII* (Ed.: A. W. Fountain), Orlando, FL, pp. C5540-C5540, 2007.
[82] C. SCHOLLHORN, A. M. FULLER, J. GRATIER, R. E. HUMMEL: Developments on standoff detection of explosive materials by differential reflectometry, Applied Optics, No.46, p. 6232-6236, 2007.
[83] R. FURSTENBERG, C. A. KENDZIORA, J. STEPNOWSKI, S. V. STEPNOWSKI, M. RAKE, M. R. PAPANTONAKIS, V. NGUYEN, G. K. HUBLER, R. A. MCGILL: Stand-off detection of trace explosives via resonant infrared photothermal imaging, Applied Physics Letters, No.93, 2008.
[84] C. W. V. NESTE, L. R. SENESAC, T. THUNDAT: Standoff Spectroscopy of Surface Adsorbed Chemicals, Analytical Chemistry, No., 2009.

Publication 2

B. Zachhuber, G. Ramer, A. Hobro and B. Lendl

Stand-off Raman Spectroscopy of Explosives

Proc. of SPIE Vol. 7838 (2010)

Stand-off Raman Spectroscopy of Explosives

Bernhard Zachhuber, Georg Ramer, Alison J. Hobro and Bernhard Lendl[1].
Vienna University of Technology, Getreidemarkt 9/164 AC, 1060 Vienna, Austria

ABSTRACT

We present our work on stand-off Raman detection of explosives and related compounds. Our system employs 532 or 355 nm laser excitation wavelengths, operating at 10 Hz with a 4.4 ns pulse length and variable pulse energy (maximum 180 mJ/pulse at 532 nm and 120 mJ/pulse at 355 nm). The Raman scattered light is collected by a co-axially aligned 6" telescope and then transferred via a fiber optic cable and spectrograph to a fast gating iCCD camera capable of gating at 500 ps. We present results including the effect of different excitation wavelengths, showing that 355 nm excitation gives rise to significantly stronger stand-off Raman signals compared to that of 532 nm. We also show the effect of appropriate detector gating widths for discrimination of ambient light and the reduction of high background signals in the obtained Raman spectra. Our system can be used to identify explosives and their precursors in both bulk and trace forms such as RDX and PETN in the low mg range and TNT in the 700 μg range at a distance of 20 m, as well as detection of a 1% or greater H_2O_2 solution at a distance of 6.3 m.

Keywords: Stand-off Raman, Explosives, Ambient light rejection, Substrate interference.

1. INTRODUCTION

New devices for the detection and identification of explosives in real scenarios must overcome a range of technical challenges. Such detection methods should be accurate, with low false positive and false negative rates, they should be adaptable to counter for the development of new explosives and should be able to detect explosives in a range of different forms such as liquids or solids as well as in bulk and trace quantities. Ideal systems should be portable, have low power consumption and should function both indoors and outdoors, in daylight, taking into account the possibility of adverse weather conditions such as rain, snow or high humidity.

A number of optical spectroscopic techniques go some way to fulfilling these criteria. The work presented here is part of a European Framework 7 project 'OPTIX: Optical Technologies for the Identification of Explosives' [1] which will combine Raman spectroscopy with laser induced breakdown spectroscopy (LIBS) and mid-infrared laser spectroscopy utilizing Quantum Cascade Lasers (QCLs). The core of the OPTIX prototype will be a pulsed Nd:YAG (1064 nm) laser capable of irradiating the sample of interest at different intensities and wavelengths using a series of harmonic generators. For stand-off Raman the sample is irradiated with a few tens of MW/cm^2. At this irradiance the sample is not destroyed and the inelastically scattered light, containing a fingerprint of the molecule, can be recorded. Upon increasing the irradiance, partial decomposition of the sample can be achieved giving rise to molecular debris such as NO and NO_2 in the case of nitro-containing substances. The ratio of NO and NO_2 is characteristic for nitro-containing explosives and shall be measured by interrogating the sample with two QCLs being selective for NO and NO_2 [2]. A further increase of the irradiating power up to a few GW/cm^2 enables LIBS spectra to be recorded. These spectra contain information on the elemental composition of the sample [3]. Overall, the OPTIX prototype aims to record orthogonal information of the object under interrogation. In this manner, the OPTIX prototype will have an increased probability of detecting a range of explosives in bulk, trace, solid and liquid forms and by exploiting the strengths of each system the sensitivity and specificity will be increased whilst making it more difficult to confuse or defeat the system.

---

[1] Bernhard.lendl@tuwien.ac.at +43(0)15880115140

Raman spectroscopy is the most generally applicable subsystem of the three, as it can be applied to any substance including both nitrogen and peroxide based explosives in liquid and solid forms. In this paper we report specifically on the achievements to date for the Raman subsystem.

2. EXPERIMENTAL PROCEDURES

2.1 Chemical Preparations and Background Substrates

NaCl, NH_4NO_3, and H_2O_2 were purchased from Sigma and used without further purification. Explosives were obtained from Entschärfungs- und Entminungsdienst (deactivation and demining service), Vienna. A number of these explosives are either pure, or can be considered to be pure (as other materials present are in a low concentration). However, for a number of the explosives pure substances were not available and, instead, commercial grade explosives were used. These were as follows:
- EGDN (ethyleneglycoldinitrate): 14.5% EGDN, 15.5% nitroglycerine, 59 % ammonium nitrate, 2.5 % glycerine, 1.2 % nitrocellulose, 5.1 % wood pulp and other burning materials, 0.7 % water.
- RDX (1,3,5-trinitroperhydro-1,3,5-triazine): Hexogene and plastic binder.
- PETN (pentaerythritol tetranitrate): 6 % plastic binder, 94 % PETN

The background substrates were glass, in the form of microscope slides, and low density polyethylene, nylon and aluminum, as 5x5 cm square blocks and sections of old cars, as 15x15 cm panels.

2.2 Stand-off Raman Instrumentation

The stand-off Raman system available at Vienna University of Technology comprises a Q-switched Nd:YAG NL301HT laser (EKSPLA, Lithuania) with a pulse length of 4.4 ns and a repetition rate of 10 Hz. The laser was operated at 532 or 355 nm excitation using doubling or tripling harmonic generators and the power was adjusted using an attenuator module (all EKSPLA, Lithuania). The laser was aligned coaxially with a 6" Schmidt-Cassegrain telescope (Celestron, USA) for the collection of Raman scattered light. The collected light passed through an appropriate long pass filter for the operating wavelength (355 or 532 nm, both from Semrock, USA) and directed to an Acton standard series SP-2785 imaging spectrograph equipped with a 300 grooves/mm grating with 500 nm blaze (Princeton Instruments, Germany) via a 19 fiber optic bundle cable (Avantes, Netherlands). Finally, the light was collected by a PI-MAX 1024RB intensified CCD (iCCD) camera (Princeton Instruments, Germany).

2.3 Fourier Transform (FT) Raman Instrumentation

In order to verify the stand-off Raman spectra obtained, FT-Raman spectra of the same substances were recorded using a Bruker IFS 66 with FT-Raman module. Raman spectra were recorded at 1064 nm laser excitation for 128 scans, taking a total time of approximately 8 minutes per spectrum.

2.4 Bulk Explosives, Precursors and Inorganic Salts Preparation

Bulk explosive measurements were preformed on the explosive samples as received with no prior preparation. For measurements of explosives on surfaces a spatula full of the bulk explosive was smeared on the background surface. The inorganic salts NaCl and NH_4NO_3 were ground to a fine powder using a pestle and a mortar. 1 g of the powder was compacted into a pellet using an oil press with pressures of 98, 157 and 235 bar applied for 2 minutes each. The resulting pellets were 10 mm in diameter and approximately 3 mm thick.

2.5 Trace Explosives Preparation

A known amount of the solid explosives (TNT, PETN, and RDX) were dissolved in acetone. The solutions were deposited on aluminum plates and measured after solvent evaporation. Amounts of each substance measured are given in the appropriate figure legends

2.6 Quantitative Measurements

For the univariate calibration NaCl/NH$_4$NO$_3$ pellets (formed from appropriate ratios of finely ground powder), placed at a stand-off distance of 9 meters, were measured using a laser power of 30 mJ/pulse and 600 coadditions taking a total time of 60 s. The constant Raman signal of nitrogen in ambient air was used to correct the measured signal intensity. For liquid measurements H$_2$O$_2$ solutions of different strengths were prepared by diluting a 30% solution with distilled water and then placed in a quartz cuvette (1 cm width, 10 mm path length). Spectra were collected using 532 nm excitation, 20 mJ/pulse for a total of 50 pulses, using 5 ns gating and a 6 mm diameter laser spot size on the target at a stand-off distance of 6.3 m.

3. RESULTS AND DISCUSSION

3.1 Effect of Measurement Parameters – Excitation Wavelength, Fluorescence Rejection and their Effect on Signal to Noise Ratio

The choice of laser excitation wavelength will have a significant impact on the performance of a stand-off Raman system. The choice often lies in a compromise between signal strength, influence of fluorescence and potential for photodegredation or dissociation of the sample. Laser excitation in the UV has the advantage of relatively high Raman signal strength and no contribution from fluorescence in the recorded Raman spectrum (as fluorescence, at UV excitation, normally lies outside the Raman spectral range) but with a high risk of photodegredation. Moving through the visible region, toward the infrared, the intensity of the Raman bands and the risk of photodegredation reduce but the interference due to fluorescence increases. By moving to the infrared, the Raman signal is usually less affected by fluorescence due to the lower excitation energies involved but consequently, data collection times must be increased, or Fourier transform instruments used, due to the lower Raman signal strength [4]. Figure 1 highlights the difference in Raman signal strength obtained with UV (335 nm) and visible (532 nm) laser excitation at 20 m, for toluene and PETN.

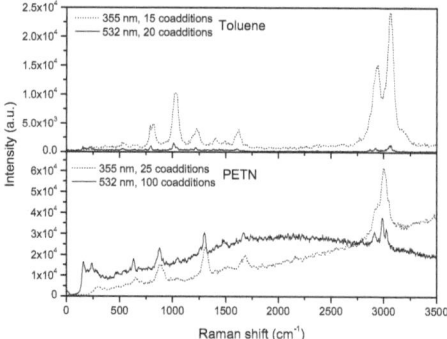

Figure 1 – Comparison of stand-off Raman spectra of Toluene (top panel) and PETN (bottom panel) obtained using 355 (dotted lines) and 532 nm (solid lines) excitation. Spectra were recorded at a distance of 20 m with a laser power of 20 mJ/pulse and 5 ns detector gating. PETN was in solid form and toluene was placed in a cuvette (1 cm width, 10 mm path length).

Figure 1 shows that, for both substances, the Raman signal obtained when using 355 nm excitation is considerably stronger than when using 532 nm excitation. The spectra of PETN also show that the signal to noise and the nature of the

Raman background also differ between the two measurements, with Raman bands originating from PETN more clearly visible in the spectrum recorded at 355 nm.

Although stand-off Raman measurements performed in the visible part of the electromagnetic spectrum can be severely affected by contributions from ambient light when working under daylight conditions, such effects can be minimized by the use of fast gating detectors in an attempt to record only the returning Raman signal and exclude light from all other sources. Figure 2 shows the effects of altering the gating speed for the stand-off detection of TNT.

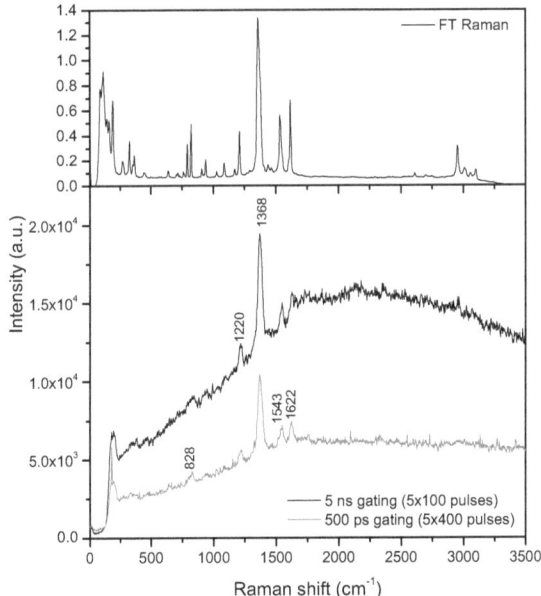

Figure 2 – FT-Raman (top panel) and stand-off Raman spectra of TNT obtained using different detector gating speeds (bottom panel). In the bottom panel the spectrum shown as a solid line was recorded with 5 ns gating and the spectrum shown as a dotted line with 500 ps gating. The number of pulses was adjusted to give rise to similar intensity Raman bands at ~1368 cm^{-1}. Both spectra were recorded at 532 nm excitation, 20 mJ/pulse and at a distance of 20 m.

When using a gating option of 5 ns, the most intense band in the TNT spectrum can be clearly observed but the lower intensity bands, especially those at higher wavenumber, are not so clearly observed and are obscured by the large background contributions recorded, as shown in Figure 2. Measuring the same sample with a 500 ps gating option results in a lower signal intensity (due to the lower exposure time) and so the number of pulses co-added for the final spectrum was increased to give rise to similar strength Raman signals as for the 5 ns gating. However, when looking at the spectrum recorded at 500 ps the most intense band is still clearly observed and the lower intensity bands at ~828, 1220, 1368, 1543 and, especially, 1622 cm^{-1} are much more clearly defined as the significant curvature to the background/baseline is reduced. The positions of the major bands in the TNT spectrum are consistent with previously published experimental data [5]. Figure 3 shows the effects of widening the gating above 5 ns.

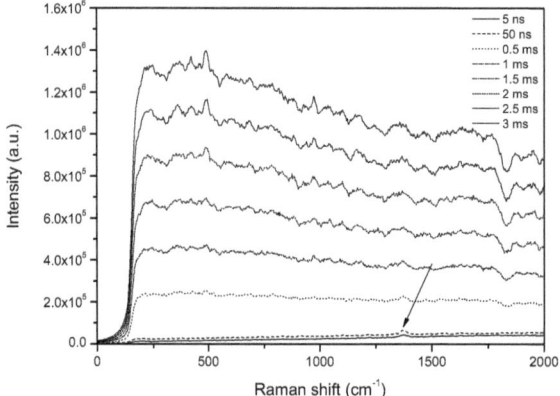

Figure 3 – Stand-off Raman spectra of TNT obtained with increasing detector gate width. Raman spectra were recorded at a distance of 20 m with a laser power of 9.97 mJ/pulse and a laser spot diameter of 6 mm at the sample. Spectra are the result of 50 co-added scans. The position of the most intense TNT band, at 1368 cm^{-1}, is shown by the arrow.

These results show that, at short detector gating widths of 5 and 50 ns, a spectrum of TNT can be observed, with relatively low background contributions to the spectrum. As the gating width increases to 0.5 ms the main TNT band can still be observed but is not so clearly defined and the background increases significantly. Increasing the gate width beyond 1 ms results in spectra with a high background and in which the characteristic peaks of TNT can no longer be distinguished.

3.2 Detection of Trace Explosives

Capability to detect small quantities is important in possible "real world" scenarios where it can be assumed that traces of explosive are left behind on objects touched after handling explosives such as bags, car doors, glasses and the like [6]. Therefore the detection of explosives on surfaces is also important and, as examples, the stand-off Raman spectra of TNT on LDPE, nylon and blue car paint are shown in Figure 4.

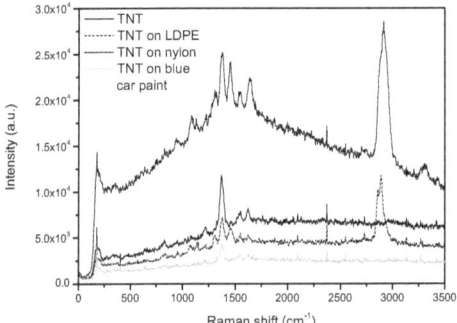

Figure 4 – Stand-off Raman spectra of pure TNT and TNT deposited on LDPE, nylon and blue car paint surfaces. Spectra were recorded at 20 m using 532 nm, 500 ps gating and 3x400 co-added pulses. The spectrum of pure TNT was recorded using a laser power of 20 mJ/pulse and a laser spot diameter of 6 mm while the TNT deposited on surfaces was measured with a laser power of 157mJ/pulse and a laser spot diameter of 25 mm.

The spectra in Figure 4 show that, when an explosive is placed on a surface with a strong Raman signal of its own, the resulting stand-off Raman spectrum obtained is an 'overlay' of the Raman spectra of the individual components. The bands arising from TNT, positioned at ~828, 1220, 1458 and 1622 cm^{-1} can be observed in all spectra. In the spectra of TNT on LDPE and nylon there are a number of Raman bands arising from the surface substance whereas the blue car paint does not give rise to any noticeable Raman bands. Only if a chemical reaction occurs between the explosive and the surface will the Raman spectra obtained change. Complications in identifying presence of an explosive through its Raman spectrum occur when marker bands for the explosive also overlap with the marker bands for the surface material. If this occurs then the limit of detection for the explosive on a particular surface may not be at the same level as for the bulk explosive.

To develop stand-off Raman as a field-deployable technique it is important to assess its capabilities for trace detection. Some example spectra of the explosives PETN, RDX and TNT at low concentrations deposited onto an aluminum surface are shown in Figure 5.

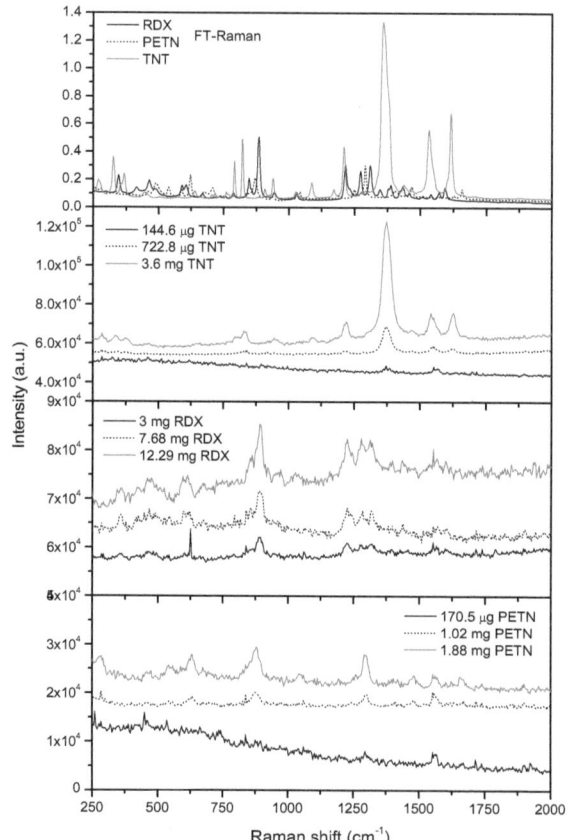

Figure 5 – Stand-off Raman spectra of low concentrations of RDX, TNT and PETN on an aluminium background. Spectra were collected using 532 nm excitation, 55.8 mJ/pulse for a total of 600 pulses, using 5 ns gating and a 10 mm diameter laser spot size on the target at a stand-off distance of 20 m.

At a relatively low laser power of 55.8 mJ/pulse, explosives such as RDX and PETN can be clearly recognized by their Raman spectral signatures at concentrations in the low mg range. Three of the bands observed in the stand-off Raman spectra of the trace PETN, positioned at ~878, 1294 and 1659 cm^{-1}, have also been observed in Raman measurements of PETN in acetone [7] and, therefore, these three bands originate from the PETN, rather than the binding material. The spectrum of RDX, with the most intense band at ~887 cm^{-1}, is also consistent with previously published spectra [5]. Once the concentration drops into the µg the Raman signatures are less well defined and unambiguous identification becomes more complex. TNT gives rise to relatively strong Raman spectra even in the upper µg range but determination becomes more difficult once the concentration is below ~150 µg.

3.3 Quantitative Stand-off Raman of Explosive Pre-cursors

Two advantages that Raman spectroscopy provides over a number of other spectroscopic detection methods is the ability to measure substances in liquid form and the ability to measure non-nitrogenous substances. We exploited these advantages for the measurement of H_2O_2 solutions as an example of an important precursor of explosives as shown in Figure 6, with the most intense band positioned at 877 cm^{-1} originating from H_2O_2 [8].

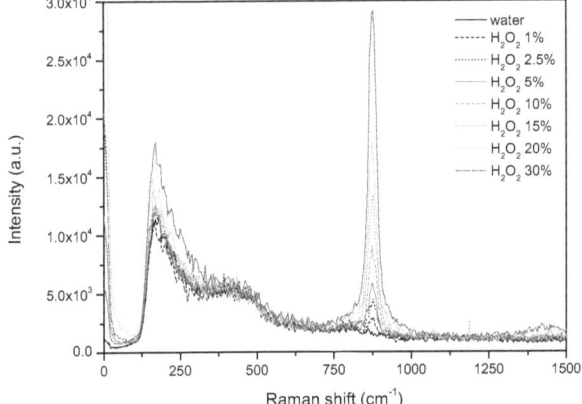

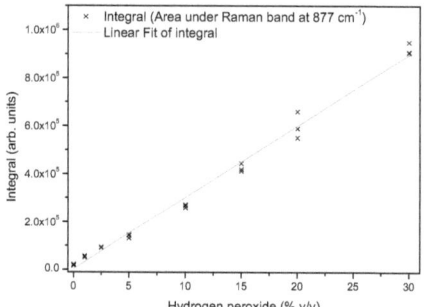

Figure 6 – Stand-off Raman spectra of different concentrations of H_2O_2 (top panel) and linear limit of detection (bottom panel). Spectra were collected using 532 nm excitation, 20 mJ/pulse for a total of 50 pulses, using 5 ns gating and a 6 mm diameter laser spot size on the target at a stand-off distance of 6.3 m. The limit of detection was calculated using the integral under the Raman band positioned at 877 cm^{-1}.

Under these conditions the measurement of H_2O_2 down to 1% is possible and a theoretical limit of detection of 0.6% was estimated.

Here we present a move towards quantitative stand-off Raman spectroscopy using an internal standard by making use of the Raman band arising from atmospheric nitrogen. The collected spectra are normalized to the intensity of the nitrogen band and in doing so, intensity fluctuations of the analyte bands, due to changes in excitation laser intensity and optical throughput, can be corrected for [9]. Raman spectra of three pellets, differing in ammonium nitrate concentration (27.76, 42.96, and 65.76 %) are shown in Figure 7.

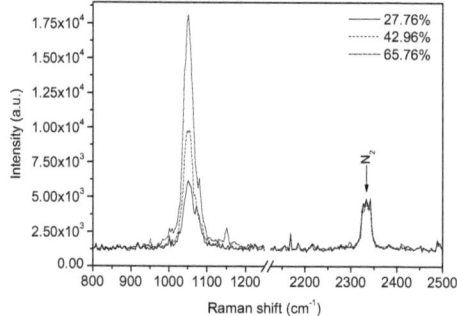

Figure 7 - Raman spectra of NH_4NO_3/NaCl pellets of different ammonium nitrate concentration (27.76, 42.96, and 65.76 %).

This is promising for future quantification in stand-off Raman spectroscopy. However, it is important to note that the ratio between the sample and nitrogen bands will change if the focus of the telescope is altered, something that may require adjustment for very strong or weak Raman scattering samples.

4. CONCLUSIONS

The results presented here show that stand-off Raman spectroscopy is capable of detecting low levels of explosives and their precursors with relatively low laser powers operating in the visible region, as exemplified by the measurements of concentrations of H_2O_2 as low as 1%. Stand-off Raman spectroscopy can also be used to detect explosives, not only in solid forms, but also traces deposited on a range of different materials, including glass, plastic and car paint, that could be encountered in real life scenarios.

The measurements presented here have been recorded using a conventional amateur astronomy telescope and, as such, is not optimized for operation at 355 nm. However, by moving to this UV wavelength we were still able to show that the Raman signal increases, as compared to excitation at 532 nm, often leading to better quality spectra in a shorter time. Future work will concentrate on the fourth harmonic of the Nd:YAG (266 nm) to further explore potential advantages in stand-off Raman resulting from deep UV excitation.

The results presented here show that such stand-off Raman systems have the potential to be developed into real world devices in the near future, adding to the growing range of 'ready to go' systems already available on the market for portable Raman detection systems already successfully employed for identification of explosives, on-site, at close distances [10].

ACKNOWLEDGEMENTS

The research leading to these results has received funding from the European Community's Seventh Framework Program (FP7/2007-2013) under Grant Agreement No 218037

REFERENCES

[1]. OPTIX. (2009) http://www.fp7-optix.eu/
[2]. Bauer, C. et al. "Potentials and limits of mid-infrared laser spectroscopy for the detection of explosives". Applied Physics B 92, 327-333(2008).
[3]. López-Moreno, C. et al. "Test of a stand-off laser-induced breakdown spectroscopy sensor for the detection of explosive residues on solid surfaces". Journal of Analytical Atomic Spectrometry 21, 55(2006).
[4]. Hobro, A.J. & Lendl, B. "Stand-off Raman spectroscopy". TrAC Trends in Analytical Chemistry 28, 1235-1242(2009).
[5]. Lewis, I.R., Daniel, N.W. & Griffiths, P.R. "Interpretation of Raman spectra of nitro-containing explosive materials. Part I: Group frequency and structural class membership". Applied spectroscopy 51, 1854-1867(1997).
[6]. Wallin, S. et al. "Laser-based standoff detection of explosives: a critical review". Analytical and bioanalytical chemistry 395, 259-74(2009).
[7]. Tuschel, D.D. et al. "Deep ultraviolet resonance Raman excitation enables explosives detection". Applied spectroscopy 64, 425-32(2010).
[8]. Venkateswaran, S. "Raman Spectrum of Hydrogen Peroxide". Nature 127, 406(1931).
[9]. Zachhuber, B., Hobro, A.J. & Lendl, B. "Quantitative stand-off Raman spectroscopy". Applied spectroscopy submitted
[10]. Moore, D.S. & Goodpaster, J.V. "Explosives analysis. Analytical and bioanalytical chemistry" 395, 245-6(2009).

Publication 3

B. Zachhuber, G. Ramer, A.J. Hobro, E. t. H. Chrysostom and B. Lendl

Stand-off Raman spectroscopy: a powerful technique for qualitative and quantitative analysis of inorganic and organic compounds including explosives

Analytical and Bioanalytical Chemistry 400 (2011)

ORIGINAL PAPER

Stand-off Raman spectroscopy: a powerful technique for qualitative and quantitative analysis of inorganic and organic compounds including explosives

Bernhard Zachhuber · Georg Ramer · Alison Hobro ·
Engelene t. H. Chrysostom · Bernhard Lendl

Received: 26 November 2010 / Revised: 21 January 2011 / Accepted: 24 January 2011 / Published online: 20 February 2011
© Springer-Verlag 2011

Abstract A pulsed stand-off Raman system has been built and optimised for the qualitative and quantitative analysis of inorganic and organic samples including explosives. The system consists of a frequency doubled Q-switched Nd: YAG laser (532 nm, 10 Hz, 4.4 ns pulse length), aligned coaxially with a 6″ Schmidt–Cassegrain telescope for the collection of Raman scattered light. The telescope was coupled via a fibre optic bundle to an Acton standard series SP-2750 spectrograph with a PI-MAX 1024RB intensified CCD camera equipped with a 500-ps gating option for detection. Gating proved to be essential for achieving high signal-to-noise ratios in the recorded stand-off Raman spectra. In some cases, gating also allowed suppression of disturbing fluorescence signals. For the first time, quantitative analysis of stand-off Raman spectra was performed using both univariate and multivariate methods of data analysis. To correct for possible variation in instrumental parameters, the nitrogen band of ambient air was used as an internal standard. For the univariate method, stand-off Raman spectra obtained at a distance of 9 m on sodium chloride pellets containing varying amounts of ammonium nitrate (0–100%) were used. For the multivariate quantification of ternary xylene mixtures (0–100%), stand-off spectra at a distance of 5 m were used. The univariate calibration of ammonium nitrate yielded R^2 values of 0.992, and the multivariate quantitative analysis yielded root mean square errors of prediction of 2.26%, 1.97% and 1.07% for o-, m- and p-xylene, respectively. Stand-off Raman spectra obtained at a distance of 10 m yielded a detection limit of 174 μg for $NaClO_3$. Furthermore, to assess the applicability of stand-off Raman spectroscopy for explosives detection in "real-world" scenarios, their detection on different background materials (nylon, polyethylene and part of a car body) and in the presence of interferents (motor oil, fuel oil and soap) at a distance of 20 m was also investigated.

Keywords Raman · Stand-off · Quantification · Remote · Explosive

Introduction

The necessity for reliable chemical information emerges from different areas of our modern society. These include environmental and process monitoring for assuring safe and efficient production, as well as biomedical diagnostics for the adoption of promising treatment strategies. Chemical information is also required in other fields such as experimental research, law enforcement and trade. To provide the requested information with the desired quality, a constant improvement in terms of sensitivity, selectivity, robustness, linear range, accuracy, precision and speed of analysis is needed. Analytical Chemistry responds to these needs by developing new instrumentation and techniques that are capable of delivering the requested chemical information with the desired quality. Often, this is accomplished by a wide range of experimental analytical techniques which, in most cases, require a dedicated laboratory to become fully operative. Examples include the combined use of separation techniques followed by different types of mass spectroscopic detection for sensitive

Published in the special issue *Analytical Sciences in Austria* with Guest Editors G. Allmaier, W. Buchberger and K. Francesconi.

B. Zachhuber · G. Ramer · A. Hobro · E. t. H. Chrysostom ·
B. Lendl (✉)
Institute of Chemical Technologies and Analytics,
Vienna University of Technology,
Getreidemarkt 9/164AC,
1060 Vienna, Austria
e-mail: bernhard.lendl@tuwien.ac.at

qualitative and quantitative analysis of liquids and gases or the use of imaging techniques for obtaining spatially resolved information. In many instances, improvements of the employed techniques were made possible by miniaturisation and integration of different operational steps of analysis. In the field of liquid phase analysis, such miniaturisation and integration led to the development of concepts such as micro-total analysis systems (μ-TAS) [1] and lab on a chip (LOC) systems [2]. Similarly, the integration of complementary imaging techniques such as atomic force microscopy with Raman scattering [3] or electrochemical detection in one single instrument also provides access to more detailed chemical information of the samples under investigation [4].

In recent years, special interest branches such as security and defence have new and unique demands for obtaining chemical information. In these areas, remote sensing is becoming extremely desirable, in particular for detection of toxic and explosive substances. Remote sensing is however not restricted to these applications as it is also requested in geological applications such mineral detection in rocks or art sample analysis. In remote sensing, the requirements are completely different in that no physical contact of the sample with the instrument is required or possible; however, the need for sensitivity, selectivity, accuracy and quantitative information remains the same. In order to bridge the gap between the sample and instrument, often laser-based optical techniques are employed, among them Raman scattering detection.

Raman scattering spectroscopy offers a non-destructive, molecular specific and quantitative means for remote detection or sensing of analytes. Its advantage lies in the fact that it yields a specific molecular fingerprint of a substance by measuring the vibrational states of a target material [5]. It should be pointed out that Raman scattering is a weak process, only one in every 10^6–10^8 scattered photon is Raman scattered [5]. The major sources of interference are due to elastically scattered light (Rayleigh scattering) and fluorescence which occurs when excited molecules relax from a higher excited state and emit light. The former process is instantaneous, whereas the latter is dependent on the lifetime of the upper energy level. By changing the laser excitation wavelength, disturbing fluorescence can be avoided. Ideally, either longer wavelengths are used [6] in order not to excite higher electronic states, or shorter wavelengths are applied [7,8] so that the fluorescence occurs in a different spectral region than the Raman signal.

Stand-off Raman scattering detection offers all the advantages of standard Raman scattering spectroscopy. The fundamental difference lies in the fact that with stand-off Raman, the sample is probed at a distance from the Raman laser excitation source. This is particularly desirable when it is necessary to minimise the risk to the operator when samples investigated are toxic, explosive or simply out of reach. Stand-off Raman spectroscopy was initially developed in the 1960s [9] and primarily used for the detection of atmospheric gases such as sulphur dioxide, nitrogen and oxygen [10]. In 1992, Angel et al. [11] extended stand-off Raman spectroscopy measurements to include inorganic compounds such as $K_4[Fe(CN)_6]$, $NaNO_2$ and $NaNO_3$, as well as organic substances like CCl_4 and acetaminophen. Sharma et al. have used this method not only for the detection of inorganic and organic compounds [12] but also for geological investigation of minerals [7] with the potential to probe samples on planets in future space missions using unmanned landing units [13–15]. Another field where remote detection is desirable is in the analysis of art and archaeological samples [16].

Over the last decade, this technique has been increasingly used for the detection of explosives. Explosives such as RDX (cyclotrimethylene trinitramine), HMX (octahydro-1,3,5,7-tetranitro-1,3,5,7-tetrazocine), TNT (trinitrotoluene), PETN (pentaerythritol tetranitrate) and TATP (triacetone triperoxide) have been detected at 50 m [17], and Semtex (which contains RDX and PETN) has been detected at 55 m [18]. In addition, urea nitrate and mixtures such as ANFO (ammonium nitrate–fuel oil) have also been investigated [19]. Ramírez-Cedeño et al. [20] have detected chemical weapon agent simulants (CWAS) such as DMMP (dimethyl methylphosphonate) in sealed glass and plastic containers at a distance of 6.7 m. Petterson et al. [21] measured 2 g TATP through a double glass window at a distance of 200 m. Furthermore, Petterson et al. [21] obtained stand-off resonant Raman spectra of TNT, DNT and nitromethane vapours at the parts per million levels using a tunable UV laser as the excitation source. A complete review of stand-off Raman detection is beyond the scope of this paper, but the reader is referred to a recent review by Hobro et al. for applications [22]. For completeness, it should be noted that there are other stand-off detection techniques in the literature and include infrared spectroscopy (IR) [23] and laser-induced breakdown spectroscopy (LIBS) [24], as well as a combination for LIBS and Raman [25].

Quantification methods for Raman spectroscopy have been available for at least 55 years. Stamm et al. [26] outlined a number of schemes to be followed for quantitative Raman analysis, covering both single and multi-component samples. More recently, a review by Pelletier [27] discussed many of the important considerations, including a comprehensive catalogue of publications associated with quantitative Raman scattering. Such quantitative Raman analyses have been used to study a whole range of polymers, hydrocarbons including xylenes, inorganic samples, as well as biological and pharmaceutical systems. The quantification used in each of these systems varies greatly, depending on the exact nature of the sample under investigation, but includes the relationship of the analyte concentration to Raman band height, area or shifts in band position, as well as more sophisticated principle component analysis, partial least squares and cross-correlation techniques [27]. The intensity of the Raman scattered light $I(v)_R$ that is generated for a particular sample at a given frequency v is given by the equation [27]:

$$I(v)_R = \frac{2^4 \pi^3}{45 \cdot 3^2 \cdot c^4} \cdot \frac{hI_L N(v_0 - v)^4}{\mu v(1 - e^{-hv/kT})}$$
$$\times \left[45(\alpha'_a)^2 + 7(\gamma'_a)^2 \right]$$

where c is the speed of light, h is Planck's constant, I_L is excitation intensity, N is the number of scattering molecules, v is molecular vibrational frequency in Hertz, v_0 is laser excitation frequency in Hertz, μ is reduced mass of the vibrating atoms, k is Boltzmann constant, T is absolute temperature, α'_a is mean value invariant of the polarisability tensor and γ'_a is anisotropy invariant of the polarisability tensor [27]. A possible alternative for quantification is the use of an internal standard present in the recorded Raman spectrum as an invariant band. Its area or height in the collected spectrum can be used for normalisation. Such an internal band can either originate from an additional material added as a standard or from a substance already present in the mixture and known to be invariant in concentration throughout the experiment [27], such as solvent bands [28]. In stand-off Raman spectrometry, bands originating from components present in ambient air such as nitrogen or oxygen can be used instead. In this paper, we present for the first time the application of univariate and multivariate calibration methods for the quantification of solid and liquid analytes from stand-off Raman spectra.

It should be noted that this work resulted as part of the EU FP7-project OPTIX [29] where the emphasis lies on the integration of Raman [30], mid-IR and LIBS in one single instrument for the remote detection of explosives, in the shortest possible time and in the presence of interferents. The explosive PETN along with ammonium nitrate (NH_4NO_3) and sodium chlorate ($NaClO_3$) were analysed as they are possible ingredients in homemade explosives due to their ease of availability as fertiliser and weed killer, respectively. We will focus here on the characterisation of our experimental setup, the measurement of pure samples, mixtures, analytes at trace levels and analytes in the presence of interferents. To demonstrate the sensitivity of our system, we have included single-shot analyses and determined the limits of detection (LOD) for our pulsed stand-off Raman system. For a complete review of laser-based methods for the trace detection of explosives, the reader is referred to the review by Wallin et al. [31].

Experimental

Chemicals

NaCl (>99.5%), $NaClO_3$ (>99%), NH_4NO_3 (>99.0%), o-xylene (98%), m-xylene (98%), p-xylene (99%) and sulphur were obtained from Sigma–Aldrich, Germany and diisopropyl methylphosphonate (DIMP 95%) from VWR, Austria. Pentaerythritol tetranitrate (PETN) was obtained from Entschärfungs- und Entminungsdienst (deactivation and demining service), Vienna. Nylon and low density polyethylene (LDPE, Riblene FL 34 LDPE Ortho Quero, Agru, Austria) and a part of a blue painted car body (10× 10 mm) were used as sample backgrounds. A 1-mm extruded polyethylene (PE) board (Wettlinger Kunststoffe, Austria) was used to evaluate the telescope field of vision. Liquid soap (olive cream soap, Hofer, Austria), motor oil and fuel oil (both OMV, Austria) were used as sample interferents.

Sample preparation

Solids

All chemicals were used without further purification. One gramme of the solid material (i.e. NaCl, $NaClO_3$ and NH_4NO_3) was compacted into pellets 10 mm in diameter using a pellet press. The inorganic salts were ground by means of an agate mortar and compacted in three steps of increasing pressure (98, 157 and 235 bar), applied for 2 min per pressure step, resulting in pellets of approximately 3 mm thickness. The plastic explosive PETN was manually pressed on different surfaces including nylon, LDPE and part of a blue painted car body. For the measurement of solid substance mixtures, 0.5 g of $NaClO_3$ was ground together with 0.5 g of NaCl in an agate mortar before the mixture was compressed into pellets as described above. For quantitative analysis, different pellets of NH_4NO_3 and NaCl were prepared. For the univariate calibration $NaCl/NH_4NO_3$ pellets, the NH_4NO_3 content of the ten sample pellets used for univariate regression ranged from 0% to 100%.

For the experiments with liquid interferents such as motor oil, fuel oil and liquid soap (30 μl) were dropped on top of the prepared $NaClO_3$ pellets forming a thin film. The pellets with the adhered liquid interferent film were then analysed. $NaClO_3$ was measured in the presence of different plastic backgrounds simply by placing the pellet in front of LDPE or nylon background.

Liquids

The liquids DIMP (diisopropyle methylphosphonate) and the three xylenes were measured in a quartz cuvette, 1 cm width and with a path length of 5 mm. The three xylenes

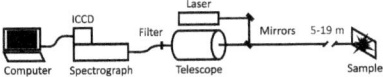

Fig. 1 Schematic representation of the pulsed stand-off Raman setup

were mixed in different ratios to produce 23 samples with differing proportions of each xylene isomer, a training set for use in building the multivariate model. A further five samples were used as a validation set to test the developed model.

Trace amounts

In order to measure $NaClO_3$ at lower levels, aqueous solutions of the inorganic salt (10 g/l) were prepared and 20–120 μl pipetted onto aluminium plates corresponding to 0.2–1.2 mg $NaClO_3$. Measurements were performed after solvent evaporation.

Confocal Raman microscope

Reference Raman spectra were obtained using a confocal Raman microscope (LabRAM, Horiba Jobin-Yvon/Dilor, Lille, France) for comparison with the stand-off Raman spectra. Raman scattering was excited by a He–Ne laser at 632.8 nm and a laser power of 7 mW. The dispersive spectrometer was equipped with a grating of 600 lines/mm, giving a spectral resolution of 4 cm^{-1}. The detector was a Peltier-cooled CCD detector (ISA, Edison, NJ, USA). The laser beam was focused manually on the sample by means of a ×20 microscope objective.

Pulsed stand-off Raman setup

The pulsed stand-off Raman system (Fig. 1) comprises a Q-switched Nd:YAG NL301HT laser (EKSPLA, Lithuania) with a pulse length of 4.4 ns and a repetition rate of 10 Hz. This excitation laser can be operated at 532 or 355 nm using doubling or tripling harmonic generator crystals. The power can be adjusted using an attenuator module (all EKSPLA, Lithuania).

The laser was aligned coaxially with a 6″ Schmidt–Cassegrain telescope (Celestron, USA) for the collection of Raman scattered light. The scattered light was filtered through an appropriate long pass filter for the relevant excitation wavelength (532 nm, 355 nm from Semrock, USA) and collected via a fibre optical bundle cable consisting of 19 200-μm diameter optical fibres (Avantes, Netherlands). The fibre optic bundle was directed to an Acton standard series SP-2750 imaging spectrograph equipped with three selectable gratings (300, 1,800 and 2,400 grooves/mm; Princeton Instruments, Germany). Finally, the light was detected by a PI MAX 1024RB intensified CCD (ICCD) camera (Princeton Instruments, Germany). The gate width of this ICCD camera can be reduced to 500 ps. During all measurements, the laser and ICCD camera were synchronised so that the measurement window coincided with the maximum Raman signal, minimising the signal contributions from fluorescence and daylight. Spectra were obtained after column wise pixel binning. For some experiments, the beam diameter was changed using a 1:4 beam expander to achieve a larger sampling area when measuring combinations of $NaClO_3$ and nylon or LDPE as backgrounds. The laser beam diameter was expanded from 6 to 25 mm, and the pulse energy increased from 20 to 157 mJ. The irradiation intensity per pulse was reduced from 71 to 32 mJ/cm^2. The excitation laser intensity was adjusted so as to avoid visible deterioration and destruction of the samples.

Results and discussion

Alignment and optimisation

Investigation of laser alignment

Good alignment of collection optics and laser beam is crucial when using telescopes with a limited field of vision, such as Schmidt–Cassegrain telescopes. In Fig. 2, the collection efficiency of the telescope is illustrated. Spectra

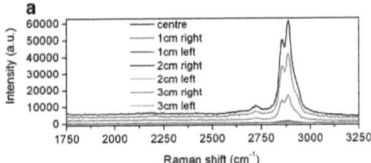

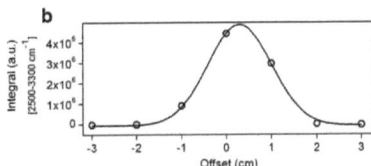

Fig. 2 Change of collection efficiency with laser alignment relative to telescope. Stand-off Raman spectra of polyethylene with different offset (**a**) recorded at a distance of 9 m using 532-nm laser pulses of 20 mJ, gate width 5 ns; 300 added spectra. **b** Baseline-corrected integral of the polyethylene bands between 2,500 and 3,300 cm^{-1} are drawn as a function of the laser offset relative from the telescope axis. A Gaussian curve was fitted through the measured data, leading to a FWHM (full width at half maximum) of 1.6 cm

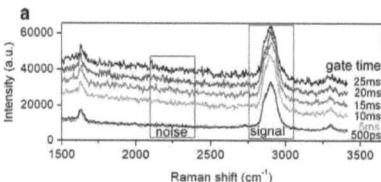

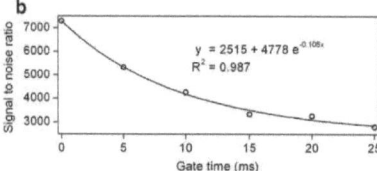

Fig. 3 Impact of the length of the gate width on the quality of the stand-off Raman spectra of nylon. **a** Spectra at a distance of 15 m; excitation with 532 nm laser; pulse energy 50 mJ/pulse; for the 500 ps gate, 100 spectra were added to compensate for shorter collection time; for all other gate widths, 10 spectra were added. **b** Signal-to-noise ratio versus gate width; band area between 2,750 and 3,050 cm^{-1}, RMS noise between 2,100 and 2,400 cm^{-1}

of a polyethylene (PE) sheet at a stand-off distance of 9 m were recorded. The laser beam was directed to different locations relative to the centre of the sample.

The zero position indicates the middle of the telescope where the collection efficiency is at its maximum. When the laser spot is placed to either side of the centre of the field of vision, the collected Raman light intensity decreases. The baseline corrected integral of the PE bands (Fig. 2a) in the spectral range from 2,500 to 3,300 cm^{-1} is shown in Fig. 2b. Fitting a Gaussian curve through the measurement points, the FWHM (full width at half maximum) was determined as 1.6 cm. This means at an angle 0.05° off the telescope axis, the light collection efficiency is reduced to 50% of the maximum.

Optimisation of gate width and time delay

A complication when measuring stand-off Raman signals is ambient light, either sunlight or artificial light. Gated signal detection synchronised with the pulsed probing laser is one of the methods used to reduce contributions due to ambient light and long-lived fluorescence. The length of the camera exposure times, i.e. the detector gate width, was investigated to see the effect on the sensitivity of the measured Raman signals. Figure 3 shows the influence of the camera exposure on Raman signals. The detector gate width was varied whilst measuring a nylon sample at a stand-off distance of 15 m. The spectra shown are the result of the summation of ten laser pulses (532 nm, pulse length 4.4 ns, energy 50 mJ per pulse, beam diameter 6 mm). As can be observed in Fig. 3a, increasing the gate width from 500 ps to 25 ms, the bands around 1,900 cm^{-1} increase along with the noise level. The signal-to-noise ratio (S/N) was calculated using the baseline corrected band area between 2,750 and 3,050 cm^{-1} (CH$_2$ stretching) and the root mean square (RMS) noise between 2,100 and 2,400 cm^{-1}. The S/N ratio with a gate of 500 ps is 7,292 and decreases to 2,807 when changing the gate to 25 ms (Fig. 3b). At a 500-ps gate width, the gate was positioned such that the most stable and intense fraction of the pulse was used to generate the Raman signal. This, in combination with a short sampling time where the noise contribution is lower, leads to a higher than expected S/N ratio at 500 ps. From this data, the necessity of short-gated, synchronised detection is apparent. In the following experiments, a 5-ns gate width was used in order to meet the constraints within the OPTIX project [29].

Reduction of fluorescence contributions can be achieved by adjusting the time delay between the excitation laser pulse and the detected Raman signal at the camera. As mentioned earlier, Raman scattering is instantaneous when compared with fluorescence, which is a prolonged process. Thus, it would be expected that by increasing the delay, the maximum Raman signal should occur at a shorter delay compared to the fluorescence background. This is indeed the case as shown in Fig. 4a where a series of spectra of DIMP with an incremental delay of 500 ps were recorded. In this experiment, the excitation laser wavelength was 355 nm, and the energy per pulse was 50 mJ. The maximum Raman

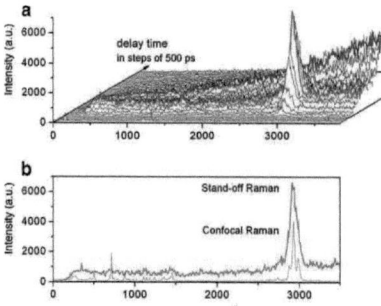

Fig. 4 Separation of Raman signal and fluorescence via short-gated, synchronised detection. **a** Five metres stand-off spectra of DIMP; ten additions of 355-nm laser pulses of 50 mJ pulse energy; fluorescence maximum (*blue*) appears 5 ns after the maximum Raman signal (*red*). **b** Maximum Raman spectrum (*red*) and confocal Raman comparison spectrum (*grey*)

signal appears 5 ns before the fluorescence. In this example, with the appropriate delay and gating settings, most of the interfering fluorescence was separated from the Raman signal. In Fig. 4b, the maximum stand-off Raman spectrum is compared to a confocal Raman spectrum of DIMP.

However, it should be noted that it is not always easy to separate the background fluorescence from the Raman signal. In particular, when soil was soaked with DIMP, the fluorescence signal due to the soil completely overwhelmed the Raman signal as shown in Fig. 5 below.

Single-shot spectra

In order to demonstrate the sensitivity of our setup, single-shot spectra of strong Raman scatterers were obtained using the shortest possible gate width of 500 ps (Fig. 6). Pure pellets of ammonium nitrate (NH_4NO_3) and sodium chlorate ($NaClO_3$) were excited by a 532-nm laser pulse of 4.4 ns length, 143 mJ pulse energy and the inelastically scattered Raman signals detected with a camera gate width of 500 ps. Figure 6 shows the spectra obtained.

These results highlight the sensitivity of the technique in that a single spectrum can be obtained during the duration of one laser pulse.

Measurement of mixtures

The Raman spectra of $NaClO_3$ in the presence of interferents (such as fuel oil, motor oil and soap) and on different backgrounds (nylon and LDPE) are shown in Fig. 7. Fuel oil, motor oil as interferents and nylon as a background material lead to an elevated baseline due to their fluorescence contribution to the signal.

In Fig. 7b, a small reduction of the $NaClO_3$ signal can be observed in the presence of soap. The spectra displayed are the summation of 100 laser pulses. In addition to the $NaClO_3$ signal, Raman bands from the plastic background appear between 1,300 and 1,500 cm^{-1} (CH_2 twisting and bending) and 2,800–3,000 cm^{-1} (CH_2 stretching). When analysing a pellet of 50% $NaClO_3$ and 50% NaCl, the bands are lower,

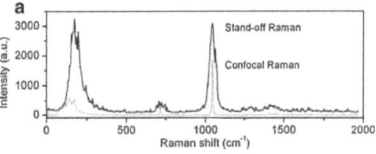

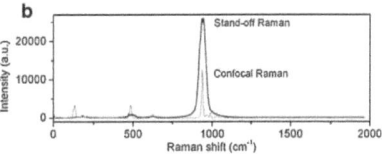

Fig. 6 Single-shot stand-off Raman spectra (*black*) of NH_4NO_3 (**a**) and $NaClO_3$ (**b**) with confocal Raman comparison spectra (*grey*). The Raman signal from one single laser pulse was recorded; camera gate time 500 ps, distance 5 m, pulse energy 143 mJ

consistent with decreased analyte concentration thus showing the potential for quantification. Characteristic band of $NaClO_3$ appears at 998cm^{-1} (ClO_3^- sym. stretching).

Figure 8 shows stand-off Raman spectra of the explosive PETN on various backgrounds including LDPE, nylon and a part of a blue car body. This clearly demonstrates the potential for the detection of explosives in real-life scenarios using stand-off Raman detection.

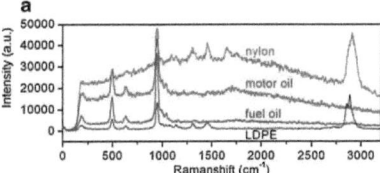

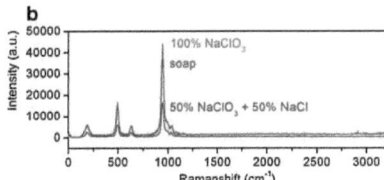

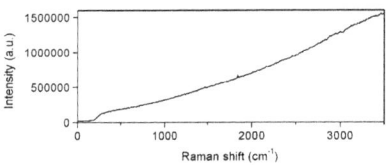

Fig. 5 Six metres stand-off spectrum of soil soaked with DIMP; 355 nm laser, 5 mJ pulse energy, 500 ps gate time, 3,600 added spectra; only fluorescence is detected

Fig. 7 a, b Nineteen metres stand-off Raman spectra of $NaClO_3$ in combination with different interferents and background materials at a distance of 19 m (532 nm, gate width 5 ns, pulse energy 20 mJ, 100 co-added laser pulses)

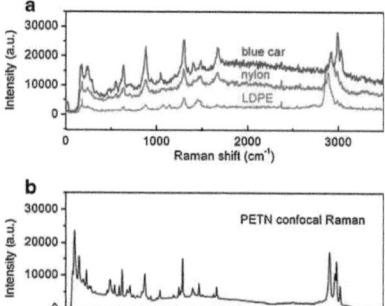

Fig. 8 PETN on different background materials. **a** Stand-off spectra at 19 m (532 nm, gate width 500 ps, pulse energy 157 mJ, 400 co-added laser pulses). **b** PETN reference spectrum

Limit of detection (LOD) determination

To evaluate the capability of stand-off Raman for trace detection at 10 m, different amounts (0.2–1.2 mg) of $NaClO_3$ were deposited on aluminium plates. Spectra from 600 laser pulses (532 nm, pulse energy 66 mJ) were co-added for each concentration sample. The baseline-corrected integrals of the most prominent $NaClO_3$ band (807–1,007 cm^{-1}, sym. stretching) were determined for each concentration. The LOD for $NaClO_3$ was determined using the new upper limit approach (ULA) where the upper confidence limit of an individual blank signal can be determined by using a critical value for the t distribution and standard error of estimate (residual standard deviation) of the regression [32]. The LOD was calculated to be 174 μg.

Quantitative stand-off Raman analysis

Univariate quantification of NH_4NO_3

For the univariate calibration experiments, $NaCl/NH_4NO_3$ pellets were placed at a stand-off distance of 9 m. A laser beam (532 nm, power of 30 mJ/pulse) was used to probe the samples. The spectra obtained per sample were the result of 600 co-additions. The NH_4NO_3 content of the ten sample pellets used for univariate regression ranged from 0% to 100%. The constant Raman signal of nitrogen in ambient air was used to correct the measured signal intensity. The validity of this method was proven by changing the light collection time from 100 to 600 pulses in steps of 100 pulses. For these measurements, the recorded signal increases with the number of pulses incorporated into each spectrum as the ICCD camera accumulates the additions additively, rather than averaging the signal. Here, a pellet of 8.2 mol% NH_4NO_3 and 91.5 mol% NaCl was measured six times for each set of pulses. For data analysis, the ammonium nitrate stretching vibration between 950 and 1,150 cm^{-1} and the N_2 band between 2,280 and 2,380 cm^{-1} were integrated after baseline subtraction.

Three representative Raman spectra of selected ammonium nitrate pellets with concentrations 21.9, 35.5 and 58.4 mol% are shown in Fig. 9a. The band at 1,050 cm^{-1} increases with ammonium nitrate concentration, whereas the nitrogen band at 2,330 cm^{-1} does not change significantly. After baseline subtraction, the band area between 950 and 1,150 cm^{-1} of different $NH_4NO_3/NaCl$ pellets and normalisation using the nitrogen Raman band area (2,280–2,380 cm^{-1}) of ambient air to correct for instrumental fluctuations, the results show a linear relation with the ammonium nitrate concentration, as depicted in Fig. 9b. The data points are averages of five measurements, and the error bars represent the 95% confidence band.

To mimic changes of instrumental settings, the exposure time was varied by changing the number of accumulated laser pulses. Using the ambient air nitrogen band to standardise measured spectra, it was possible to change the data collection time from 100 up to 600 added laser pulses without changing the resulting NH_4NO_3/N_2 band ratio, as depicted in Fig. 10.

Both the ammonium nitrate and the nitrogen signal increase with longer exposure times. However, the ratio of

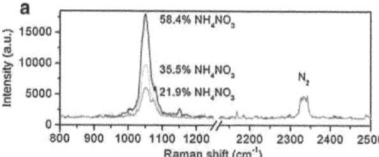

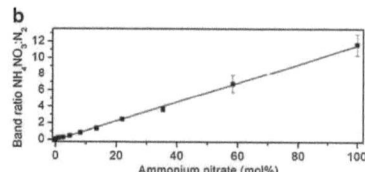

Fig. 9 a Selected Raman spectra of $NH_4NO_3/NaCl$ pellets of different ammonium nitrate concentration (21.9, 35.5 and 58.4 mol%). **b** Calibration corrected with ambient air nitrogen band; the band ratio of NH_4NO_3 and N_2 is shown. $y=0.116x-0.0973$, $R^2=0.998$

the two signals is independent of instrumental parameters such as laser power, number of accumulations, facilitating the long-term comparability of recorded spectra.

Multivariate quatification of xylenes isomers

For the multivariate calibration, the xylene mixtures were placed at a distance of 5 m from the setup in a 5-mm path length quartz cuvette and measured using a laser power of 180 mJ/pulse. Six hundred laser pulses were added per measurement. The multivariate xylene calibration was performed using OPUS 6.5 (Bruker, Germany) and the Quant 2 method. The spectral range was restricted to 850.44–796.35 and 760.45–708.33 cm^{-1}, covering the two regions where characteristic Raman bands were observed for each of the xylene samples. The training set with the 23 different mixtures of the three xylenes ranges from 0% to 100% individual xylene content. Two of these mixtures were identified as outliers and not included in the calibration model. All other training set spectra were incorporated into the model without any pre-processing steps. Five additional xylene mixtures were prepared as validation samples (differing in individual xylene content from those included in the training set). The results of this multivariate calibration and validation process are reported in terms of the coefficient of determination (R^2), which reports the percentage variance of the true component

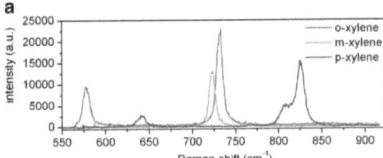

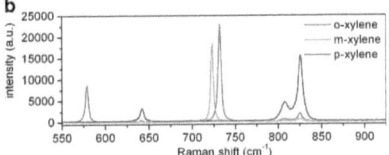

Fig. 11 **a** Stand-off Raman spectra of 100% *o*-, *m*- and *p*-xylene. **b** Confocal Raman comparison spectra

values, the root mean square error of estimation (RMSEE) and the root mean square error of prediction (RMSEP) as a criterion to judge the quality of the prediction method.

The stand-off Raman spectra obtained for the three pure xylenes, in the range 557–923 cm^{-1}, are shown in Fig. 11, and the corresponding R^2, RMSEP and RMSEE values for the multivariate calibration are shown in Table 1.

In both the training set and the validation set, the *p*-xylene shows the best fit to the calibration model with R^2 values of 0.9989 and 0.9971, respectively. The second measure of how well the calibration model explains the variance in the dataset is the RMSEE; the closer the value is to zero, the better the model describes the variability in the observations. The RMSEE, at 1.07 for *p*-xylene, is significantly smaller than that of the other two xylenes, and this is probably due to the fact that the main Raman band in the *p*-xylene spectrum does not overlap with bands from the other two xylenes. This is in agreement with another Raman study of xylenes in the presence of other petroleum constituents [33]. In summary, the R^2 values for all three xylenes in both the training and validation sets are above 98.5, indicating less than 1.5%

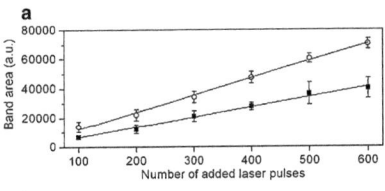

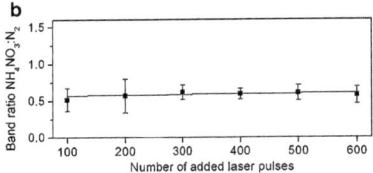

Fig. 10 **a** Band area of 8.2 mol% ammonium nitrate pellet (between 950 and 1,150 cm^{-1}, *black squares*) and nitrogen (between 2,280 and 2,380 cm^{-1}, *white circles*) with changing exposure time; average of six measurements; *error bars* show 95% confidence band. **b** The ratio of band area of NH$_4$NO$_3$ and N$_2$ shows no significant dependency with the number of added laser pulses. $Y_{NH4NO3/N2} = (0.56\pm0.14)+(7\times10^{-5}\pm3\times10^{-4})X$, $R^2=0.009$

Table 1 R^2, RMSEE and RMSEP values for the multivariate calibration of *o*-, *m*- and *p*-xylene

Xylene	Training set		Validation set	
	R^2	RMSEE	R^2	RMSEP
o	0.9972	1.80	0.9965	2.26
m	0.9961	1.97	0.9952	1.97
p	0.9989	1.07	0.9971	1.07

variance between the predicted concentration values and the true concentration values, while for m- and p-xylene, this improves to less than 0.5% variance.

Conclusion

The importance to meet the additional requirements of pulsed stand-off Raman spectroscopy was exemplified. The importance of careful laser alignment was demonstrated. The strong influence of the camera gate on the spectral quality was observed, and the need for fast-gated detection systems, in order to suppress the effect of ambient light for real-world applications, was highlighted. It was also shown that an appropriately chosen delay between laser pulse and detection can separate Raman signals from disturbing fluorescence. Using the optimised system, we demonstrated the capability of our pulsed stand-off Raman system to detect $NaClO_3$ with a LOD of 174 µg on aluminium at a distance of 10 m. In addition, it was possible to identify $NaClO_3$ in the presence of several interferents and backgrounds, thus simulating scenarios outside of the laboratory.

The application of quantitative analysis techniques to stand-off Raman spectra was demonstrated. For the first time, a univariate calibration of NH_4NO_3 in a NaCl matrix was achieved, showing that such a quantitative analysis is effective for stand-off Raman spectra obtained from solid samples. Constant nitrogen in ambient air was used to compensate for instrumental fluctuations. A multivariate calibration of a three-component xylene mixture has shown that quantitative analysis of stand-off Raman spectra of liquid multicomponent mixtures is also achievable. The validation of this model indicates for o-xylene prediction of concentration is accurate to ±1.5% and to ±0.5% for m- and p-xylene.

Acknowledgements The research leading to these results has received funding from the European Community's Seventh Framework Program (FP7/2007-2013) under Grant Agreement No. 218037 and from the Austrian Research Promotion Agency (FFG) under the Research Studios Austria program.

References

1. Manz A, Harrison JD, Verpoorte EMJ, Fettinger JC, Luedi H, Widmer HM (1991) Chimia 45:103
2. Ramsey JM (1999) Nat Biotechnol 17:1061
3. Stöckle RM, Suh YD, Deckert V, Zenobi R (1999) Chem Phys Lett 318:131
4. Wang L, Kowalik J, Mizaikoff B, Kranz C (2010) Anal Chem 82:3139
5. Smith E, Dent G (2005) Modern Raman spectroscopy a practical approach. Wiley, Chichester
6. Lewis IR, Daniel NW Jr, Chaffin NC, Griffiths PR, Tungol MW (1995) Spectrochim Acta Part A Mol Biomol Spectrosc 51:12
7. Sharma SK, Lucey PG, Ghosh M, Hubble HW, Horton KA (2003) Spectrochim Acta Part A Mol Biomol Spectrosc 59:2391
8. Misra AK, Sharma SK, Chio CH, Lucey PG, Lienert B (2005) Spectrochim Acta Part A Mol Biomol Spectrosc 61:2281
9. Cooney J (1965) Proceedings of the symposium on electromagnetic sensing of the earth from satellites. Polytechnic P, Brooklyn, New York
10. Leonard DA (1967) Nature 216:142
11. Angel SM, Kulp TJ, Vess TM (1992) Appl Spectrosc 46:1085
12. Sharma SK, Misra AK, Lucey PG, Angel SM, McKay CP (2006) Appl Spectrosc 60:871
13. Sharma SK, Misra AK, Singh UN (2008) Proc of SPIE 7153:715307-1
14. Sharma SK, Misra AK, Clegg SM, Barefield JE, Wiens RC, Acosta T (2010) Phil Trans R Soc A 68:3167
15. Klein V, Popp J, Tarcea N, Schmitt M, Kiefer W, Hofer S, Stuffler T, Hilchenbach M, Doyle D, Dieckmann M (2004) J Raman Spectrosc 35:433
16. Vandenabeele P, Castro K, Hargreaves M, Moens L, Madariaga JM, Edwards HGM (2007) Anal Chim Acta 588:108
17. Thorley FC, Baldwin KJ, Lee DC, Batchelder DN (2006) J Raman Spectrosc 37:335
18. Pettersson A, Johansson I, Wallin S, Nordberg H, Östmark H (2009) Propellants Explos Pyrotech 34:297
19. Gaft M, Nagi L (2008) Opt Mater 30:1739
20. Ramírez-Cedeño ML, Ortiz-Rivera W, Pacheco-Londoño LC, Hernández-Rivera SP (2010) IEEE Sens J 10:693
21. Pettersson A, Wallin S, Östmark H, Ehlerding A, Johansson I, Nordberg M, Ellis H, Al-Khalili A (2010) Proc SPIE 7664:76641K-1
22. Hobro AJ, Lendl B (2009) Trends Anal Chem 28:1235
23. Bauer C, Sharma AK, Willer U, Burgmeier J, Braunschweig B, Schade W, Blaser S, Hvozdara L, Müller A, Holl G (2008) Appl Phys B: Lasers Opt 92:327
24. Mordmueller M, Bohling C, John A, Schade W (2009) Proc SPIE 7484:74840F
25. Moros J, Lorenzo JA, Lucena P, Tobaria LM, Laserna JJ (2010) Anal Chem 82:1389
26. Stamm RF (1945) Anal Chem 17:318
27. Pelletier MJ (2003) Appl Spectrosc 57:20A
28. Aarnoutse PJ, Westerhuis JA (2005) Anal Chem 77:1228
29. EU FP7-project OPTIX. Available at http://www.fp7-optix.eu. Accessed 9 Nov 2010
30. Zachhuber B, Ramer G, Hobro AJ, Lendl B (2010) Proc SPIE 7838:78380F
31. Wallin S, Pettersson A, Östmark H, Hobro A (2009) Anal Bioanal Chem 395:259
32. Mocak J, Bond AM, Mitchell S, Scollary G (1997) Pure Appl Chem 69:297
33. Cooper JB, Flecher PE, Vess TM, Welch WT (1995) Appl Spectrosc 49:586

Publication 4

B. Zachhuber, C. Gasser, A. Hobro, E. t. H. Chrysostom, B. Lendl

Stand-off spatial offset Raman spectroscopy – A distant look behind the scenes

Proc. of SPIE Vol. 8189 (2011)

Stand-off spatial offset Raman spectroscopy – A distant look behind the scenes

Bernhard Zachhuber, Christoph Gasser, Alison J. Hobro, Engelene t. H. Chrysostom, Bernhard Lendl*

Vienna University of Technology, Getreidemarkt 9/164 AC, 1060 Vienna, Austria

ABSTRACT

A pulsed (4.4 ns pulse length) frequency doubled Nd:YAG laser, operating at 10 Hz, was used to generate Raman scattering from samples at a distance of 12 m. The scattered light was collected by a 6 inch telescope and the Raman spectrum recorded using an Acton SP-2750 spectrograph coupled to a gated ICCD detector. To extend the potential applications further, employing a spatial offset between the point where the laser hit the sample and the focus of the telescope on the sample, enabled collection of Raman photons that were predominantly generated inside the sample and not from its surface. This is especially effective when the content of concealed objects should be analysed. Raman spectra of H_2O_2 in a 1.5 mm thick, fluorescent HDPE plastic bottle were recorded at a distance of 12 m. From the recorded spectra it was possible to determine the H_2O_2 concentration in the concentration range from 2-30%. Stand-off Raman spectra of eleven potentially dangerous chemicals (commercial and improvised explosives) were recorded at a distance of 100 m.

Keywords: stand-off, remote, explosives, Raman, spatial offset, SORS, warfare agent simulant, quantification

*bernhard.lendl@tuwien.ac.at; phone +43 1 58801-15140; fax +43 1 58801-15199

1. INTRODUCTION

The growing number of scientists developing methods for the detection and identification of explosives and war agents reflects the growing demands for such techniques. The employed approaches are as versatile as the challenging scenarios they have to cope with. While the range of different techniques is too numerous to discuss in detail two applications of note are the detection of triacetone triperoxide, an increasingly important explosive in Improvised Explosive Devices (IEDs), using ion mobility spectroscopy[1] and the use of laser electrospray mass spectrometry for the identification of explosive formulations[2]. A broader overview of the various explosives detectors and their principles of operation was presented by Yinon et al.[3].

In contrast to chromatographic methods, which rely on close interaction with the analyte, optical methods can be successfully employed from a distance. This is especially advantageous when dangerous samples are investigated such as explosives, toxic materials or even war agents and nerve gases. An overview about stand-off detection for explosives has been given by Wallin et al.[4]. Raman spectroscopy is an optical technique widely used for the detection of explosives, both in close contact and stand-off configurations[5, 6, 7]. However, stand-off Raman spectroscopy is not limited to explosive detection. It is generally useful when it is difficult to directly reach sample material. Minerals where identified at long distances[8] in preparation for Mars missions, and even icebergs can be analysed[9] from the distance of 120 m.

Another way to detect samples from a distance is Laser Induced Breakdown Spectroscopy (LIBS)[10]. The combination of Raman spectroscopy and LIBS leads to even more robust detection of distant objects[11].

However, when the content of a container should be analysed, the success of most optical techniques relies on a relatively clear line of sight to the sample. To overcome this limitation, at least for very short sampling distances, Matousek et al. developed a technique called Spatial Offset Raman Spectroscopy (SORS)[12]. This method uses the spatial broadening of laser irradiation in a turbid medium due to scattering. By detecting the Raman signal at a sample position different from the point where the excitation laser enters the sample, it is possible to record deeper sample layers rather than the top layer or surface of the sample.

The work presented here describes the combination of stand-off Raman spectroscopy and SORS, enabling the determination of container contents at a distance of 12 m.

To evaluate the change of collected Raman signal intensity with increasing distance between sample and setup, a plastic sheet was placed at various distances ranging from 10 to 100 metres. Furthermore, eleven samples (explosives and a warfare agent simulant) were analysed at a distance of 100 metres.

2. EXPERIMENTAL PROCEDURE

2.1 Chemicals

Sulfur, $NaClO_3$, $KClO_3$, NH_4NO_3, DIMP (diisopropyl methylphosphonate) and H_2O_2 were purchased from Sigma and used without further purification. The following explosives were obtained from the Austrian Armed Forces (Armament and Defence Technology Agency; explosives, materials and POL division): ANFO (ammonium nitrate fuel oil), TNT (trinitrotoluene), RDX (cyclotrimethylenetrinitramine), PETN (pentaeritrol tetranitrate), octogen and TATP (triacetone triperoxide). PE (polyethylene) was purchased from AGRU Kunststofftechnik GmbH, Austria.

2.2 Stand-off Raman Instrumentation

The stand-off Raman system used here comprises a Q-switched Nd:YAG NL301HT laser (EKSPLA, Lithuania) with a pulse length of 4.4 ns and a repetition rate of 10 Hz. The laser operated at 532 nm via a doubling harmonic generator and the power was adjusted using an attenuator module (all EKSPLA, Lithuania). The laser was aligned coaxially with a 6'' Schmidt-Cassegrain telescope (Celestron, USA) for the collection of Raman scattered light. The collected light then passed through a long pass filter appropriate for the operating wavelength of 532 nm (from Semrock, USA) and directed to an Acton standard series SP-2750 imaging spectrograph equipped with two gratings with 300 and 1800 grooves/mm (Princeton Instruments, Germany) via a 19 fibre optic bundle cable (Avantes, Netherlands). Finally, the light was detected by a PI-MAX 1024RB intensified CCD (ICCD) camera (Princeton Instruments, Germany).

2.3 Confocal Raman Microscope

In order to verify the stand-off Raman spectra, conventional Raman spectra of the same substances were recorded using a confocal Raman microscope (LabRAM system, Jobin-Yvon/Dilor, Lille, France). Raman scattering was excited by a He-Ne laser at 632.8 nm. The dispersive spectrometer was equipped with a grating of 600 lines/mm. The detector was a Peltier-cooled CCD detector (ISA, Edison, NJ).

3. RESULTS AND DISCUSSION

3.1 Influence of Raman stand-off distance on signal intensity

In order to investigate the decrease of collected Raman signal with increasing distance from the sample, a 25 mm thick poly ethylene (PE) sheet was used as sample material. The sample was placed at different stand-off distances ranging from 10 to 100 m in steps of 10 m. At distances from 10 to 40 m the laser output pulse energy was set to 50 mJ in order to avoid sample degradation. However, to compensate for atmospheric losses the pulse energy was increased to 223 mJ for distances from 50 to 100 m. For each distance the measurement system was optimised for maximum signal collection efficiency. Spectra recorded at each stand-off distance were a result of 600 co-added laser pulses. In Figure 1 the resulting PE spectra at different stand-off distances are shown.

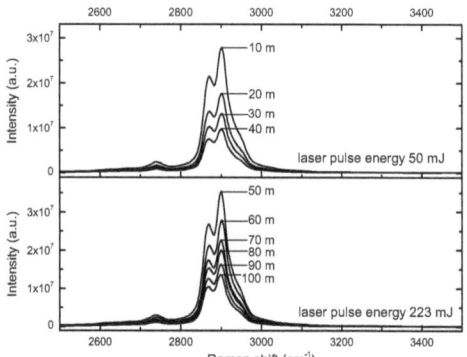

Figure 1. Reduction in intensity of the collected Raman signal with increasing stand-off distance; Raman spectra of 25 mm-PE sheet at different distances, stated in the diagram; laser pulse energy: top 50 mJ, bottom 223 mJ

The top panel of Figure 1 shows the PE spectra measured between 10 and 40 m stand-off distance, whereas the bottom panel summarises the data recorded between 50 and 100 m. As Raman scattered light does not leave excited matter in a specific direction, the number of photons entering the telescope, decreases with the distance from the sample. In order to establish a quantitative relationship between the distance and recorded signal, the integral of the PE band observed between 2400 and 3400 cm^{-1} was calculated and a baseline subtracted. Taking into account the linear increase of Raman signal with higher excitation energy the determined integrals were divided by the laser pulse energy measured at the laser head in millijoule. The resulting values are shown in Figure 2.

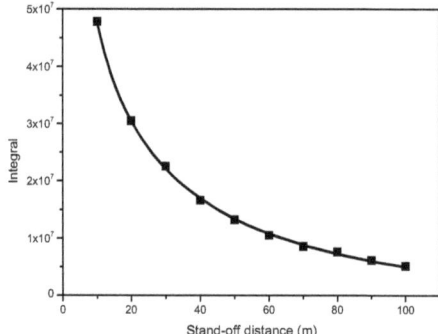

Figure 2. Variance of the PE Raman signal with stand-off distance; data points (black squares) are the baseline corrected PE intensity integrals (2400-3400 cm^{-1}) divided by pulse energy in mJ; regression line (black line): Integral=1.57E8/distance^0.46*exp(-0.0134*distance), adjusted R^2=0.9995

Two effects affect the decrease in Raman intensity at longer distances. Firstly, the reduction of collected Raman signal with increasing distance can be explained geometrically. The sample irradiates Raman scattered light in all directions, therefore, only a fraction of light can be collected via the telescope. This geometric effect results in a reduction in intensity with the square of the distance, as the imaginary surface of the spreading light sphere increases with the square of the radius, whereas the collection area is constant. The exponential term (exp(-k*distance)) was added to the equation (Equation 1) to compensate for the atmospheric light reduction[13]. The laser power reaching the sample surface decreases with increasing distance, as air molecules, as well as airborne dust particles scatter the photons elastically and inelastically. Equation 1 summarises the key factors that influence the Raman signal I at a particular stand-off distance r. (regression line Figure 2).

$$I = H \frac{1}{r^p} \exp(-k \cdot r)$$

Equation 1: Raman intensity I at different stand-off distances r; k attenuation coefficient of the atmosphere, p distance proportionality, H includes setup constants such as telescope area and substance specific Raman cross section

3.2 Stand-off Raman spectra of different explosives

Eleven substances were analysed at a stand-off distance of 100 m. Explosives (such as ANFO, TNT, RDX, PETN and octogen) used in industry as well as for military applications were analysed. Furthermore, potential explosive precursors such as sodium- and potassium chlorate ($NaClO_3$, $KClO_3$) as well as ammonium nitrate (NH_4NO_3) were also investigated, due to their general availability as weed killers or fertiliser. Triaceton triperoxide (TATP) was detected as a highly relevant substance used in Improvised Explosive Devices (IEDs). In addition spectra of DIMP (diisopropyl methylphosphonate) were recorded as this liquid is a chemical warfare agent simulant. Each substance was contained in a round glass vial (diameter 15 mm, height 20 mm) and placed 100 m from the apparatus. A representative spectrum of TATP is shown in Figure 3, together with the reference spectrum, recorded on a confocal Raman microscope.

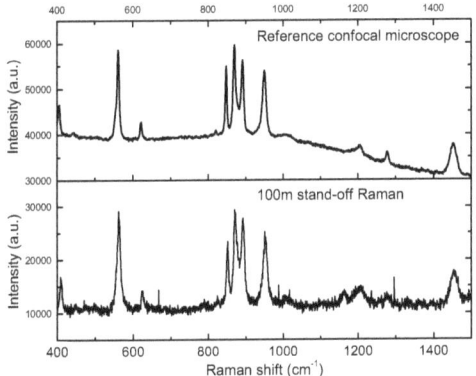

Figure 3. Raman spectra of TATP; bottom: stand-off spectrum at 100 m; top: reference spectrum measured via confocal microscope

The spectral features of TATP, depicted in the 100 m-stand-off spectrum (Figure 3, bottom) clearly correspond to the bands of the reference spectrum.

Generally challenging for Raman spectroscopy is the detection of fluorescent samples as the strong fluorescence can overlay the Raman signals. One way to reduce the amount of fluorescence detected is the use of short pico-second laser pulses and accordingly short camera gating times. In this way the instantaneous Raman photons enter the detection system prior to fluorescence photons from electronic states with longer lifetime. Appropriate timing between the 4.4 ns laser excitation pulses and camera detection gate allowed separation of Raman and fluorescence of DIMP[14]. Even more effective, this principle was applied to several substances by Åkeson et al. using 10 ps laser pulses[15]. A technique able to separate the signal from top and deeper sample layers is called Spatial Offset Raman Spectroscopy (SORS)[16]. The principle of this method is the spatial broadening of the excitation laser in a turbid sample due to scattering. Whereas in conventional Raman spectroscopy the point of laser excitation on the sample is identical with the point of detection, with SORS these two spots are spatially separated. With increasing offset between these two points the relative amount of photons from deeper layers increases. Therefore, it is possible to distinguish whether an analyte is located at the sample surface or situated in a deeper layer. In this work, stand-off Raman spectroscopy was combined with SORS allowing content detection in opaque or even white plastic bottles.

3.3 Stand-off Spatial Offset Raman Quantification

A turbid HDPE (high density polyethylene) bottle with a wall thickness of 1.5 mm was filled with a 25% H_2O_2 solution. In order to show the power of stand-off SORS the plastic container was coated with a layer of motor oil to increase the level of fluorescence obtained. The effect of changed spatial offset is shown in Figure 4.

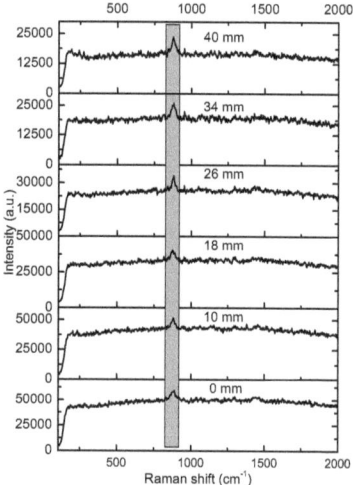

Figure 4. Spatial offset stand-off Raman spectra of H_2O_2 in a fluorescent, opaque HDPE bottle at 12 m distance; the spatial offset between laser excitation point and detector position is noted in the figure key for each panel; the grey box highlights the Raman band positioned at 877 cm^{-1}

Whereas the stand-off spectrum without offset (bottom Figure 4) shows an elevated fluorescence background, this baseline gets lower with increasing offset. The change in baseline corrected band height at 877 cm^{-1} is depicted more clearly in Figure 5 (a).

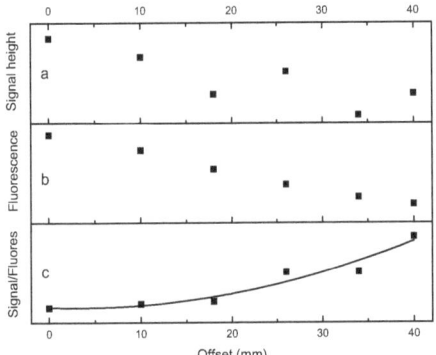

Figure 5. Influence of spatial offset; a: signal decrease of container content (hydrogen peroxide) with increasing offset; b: fluorescence background from oily plastic container decreases with increasing offset as well; c: ratio of H_2O_2-signal and fluorescence increases with increasing offset, facilitating detection of the container content

In Figure 5 (a) the decrease of analyte signal decreases with spatial offset. However, the same is true for the fluorescence background (b). Since the disturbing fluorescence baseline is caused by the top layer of the probed container, it decreases significantly earlier than the content signal. Therefore, the ratio of container content and fluorescence (c) is greater at higher offsets, providing emphasised spectral quality for the peroxide in the bottle. Taking advantage of the lower fluorescence from the container, different concentrations of hydrogen peroxide (2-30%) were measured. The linear relationship between concentration and band intensity is shown in Figure 6.

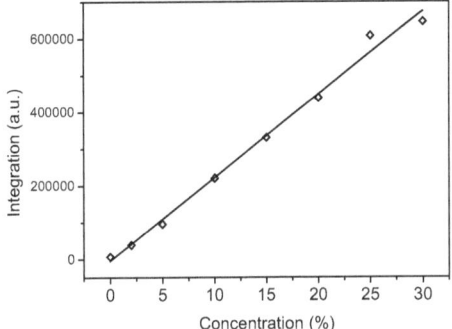

Figure 6. Stand-off spatial offset Raman quantification of hydrogen peroxide in a fluorescent, opaque HDPE plastic bottle at a distance of 12 metres

4. CONCLUSION

In this work 100 m-stand-off Raman detection of eleven security relevant substances is described. Professional explosives such as ANFO, TNT, RDX, PETN and octogen were recorded. In addition, substances used in Improvised Explosive Devices (IEDs) such as potassium- and sodium chlorate as well as ammonium nitrate and triacetone triperoxide were analysed. Furthermore, the reduction of collected Raman signal with increasing sampling distance was investigated. Most importantly, the combination of stand-off Raman spectroscopy and Spatial Offset Raman Spectroscopy (SORS) was introduced as a novel method for the detection of concealed content in distant objects. As an excample, aqueous solutions of different hydrogen peroxide concentration, were analysed in a fluorescent, opaque plastic bottle at a distance of 12 metres.

ACKNOWLEDGEMENT

The research leading to these results has received funding from the European Community's Seventh Framework Program (FP7/2007-2013) under Grant Agreement No. 218037

LITERATURE

[1] Buttigieg, G. A., Knight, A. K., Denson, S., Pommier, C. and Denton, M. B., "Characterization of the explosive triacetone triperoxide and detection by ion mobility spectrometry" Forensic Sci. Int. 135, 53–59 (2003).
[2] Brady, J. J., Judge, E. J. and Levis, R. J., "Identification of explosives and explosive formulations using laser electrospray mass spectrometry" Rapid Commun. Mass Spectrom. 24, 1659–1664 (2010).
[3] Yinon, J., "Field detection and monitoring of explosives" Trends Anal. Chem. 21, 292-301 (2002).
[4] Wallin, S., Pettersson, A., Östmark, H., Hobro, A., Zachhuber, B., Lendl, B., Mordmüller, M., Bauer, C., Schade, W., Willer, U., Laserna, J.J. and Lucena, P., "Standoff trace detection of explosives – a review" New Trends Res. Energ. Mater., Proc. Semin., 12th, (2009).
[5] Pettersson, A., Wallin, S., Östmark, H., Ehlerding, A., Johansson, I., Nordberg, M., Ellis, H. and Al-Khalili, A., "Explosives standoff detection using Raman spectroscopy: From bulk towards trace detection" Proc. of SPIE 7664, 1-12 (2010).
[6] Zachhuber, B., Ramer, G., Hobro, A. J. and Lendl, B., "Stand-off Raman spectroscopy of explosives" Proc. of SPIE 7838, 78380F (2010).
[7] Waterbury, R. D., Ford, A. R., Rose, J. B. and Dottery, E. L., "Results of a UV TEPS-Raman energetic detection system (TREDS-2) for standoff detection" Proc. of SPIE 7304, 73041B (2009).
[8] Sharma, S. K., Misra, A. K., Clegg, S. M., Barefield, J. E., Wiens, R. C. and Acosta, T., "Time-resolved remote Raman study of minerals under supercritical CO_2 and high temperatures relevant to Venus exploration" Phil. Trans. R. Soc. A 368, 3167-3191 (2010).
[9] Rull, F., Vegas, A., Sansano, A. and Sobron, P., "Analysis of Arctic ices by Remote Raman Spectroscopy" Spectrochim. Acta, Part A 80, 148–155 (2011).
[10] Sallé, B., Mauchien, P. and Maurice, S., "Laser-Induced Breakdown Spectroscopy in open-path configuration for the analysis of distant objects" Spectrochim. Acta, Part B 62, 739–768 (2007).
[11] Moros, J. and Laserna, J. J., "New Raman-Laser-Induced Breakdown Spectroscopy Identity of Explosives Using Parametric Data Fusion on an Integrated Sensing Platform" Anal. Chem., 83, 6275–6285 (2011).
[12] Matousek, P, Morris, M. D., Everall, N., Clark, I. P., Towrie, M., Draper, E., Goodship, A. and Parker, A. W., "Numerical Simulations of Subsurface Probing in Diffusely Scattering Media Using Spatially Offset Raman Spectroscopy" Appl. Spectrosc. 59, 1485-1492 (2005).
[13] Killinger, D. K., [LIDAR and Laser Remote Sensing], Handbook of vibrational spectroscopy, Ed. J. Chalmers and P. Griffiths, John Willey & Sons Ltd, Chichester, 2002.
[14] Zachhuber, B., Ramer, G., Hobro, A. J., Chrysostom, E. t. H. and Lendl, B., "Stand-off Raman spectroscopy: a powerful technique for qualitative and quantitative analysis of inorganic and organic compounds including explosives" Anal. Bioanal. Chem. 400, 2439-2447 (2011).
[15] Åkeson, M., Nordberg, M., Ehlerding, A., Nilsson, L. E., Östmark, H. and Strömbeck, P., "Picosecond laser pulses improves sensitivity in standoff explosive detection" Proc. of SPIE 8017, 80171C-80171C-8 (2011).

[16] Matousek, P., Clark, I. P., Draper, E. R. C., Morris, M. D., Goodship, A. E., Everall, N., Towrie, M., Finney, W. F. and Parker, A. W., "Subsurface probing in diffusely scattering media using spatially offset Raman spectroscopy" Appl. Spectrosc. 59, 393-400 (2005).

Publication 5

B. *Zachhuber*, C. Gasser, E. t. H. Chrysostom and B. Lendl

Stand-Off Spatial Offset Raman Spectroscopy for the Detection of Concealed Content in Distant Objects

Analytical chemistry 83 (2011)

Stand-Off Spatial Offset Raman Spectroscopy for the Detection of Concealed Content in Distant Objects

Bernhard Zachhuber, Christoph Gasser, Engelene t.H. Chrysostom, and Bernhard Lendl*

Institute of Chemical Technologies and Analytics, Vienna University of Technology, Getreidemarkt 9/164AC, 1060 Vienna, Austria

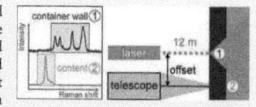

ABSTRACT: A pulsed (4.4 ns pulse length) frequency-doubled Nd:YAG laser operated at 10 Hz was used to generate Raman scattering of samples at a distance of 12 m. The scattered light was collected by a 6 in. telescope, and the Raman spectrum was recorded using an Acton SP-2750 spectrograph coupled to a gated intensified charge-coupled device (ICCD) detector. Applying a spatial offset between the point where the laser hit the sample and the focus of the telescope on the sample enabled collection of Raman photons that were predominantly generated inside the sample and not from its surface. This is especially effective when the content of concealed objects should be analyzed. High-quality Raman spectra could be recorded, within 10 s of data acquisition, from a solid ($NaClO_3$) as well as a liquid (isopropyl alcohol) placed inside a 1.5 mm thick opaque low-density polyethylene (LDPE) plastic bottle. The applied spatial offset was also advantageous in cases where the surface of the container was highly fluorescent. In such a situation, Raman spectra of the sample could not be recorded when the sampling volume (telescope observation field) coincided with the focus of the excitation laser. However, with the use of a spatial offset of some millimeters, a clear Raman spectrum of the content (isopropyl alcohol) in a strongly fluorescent plastic container was obtained.

A nalytical chemistry seeks to develop instrumental techniques for obtaining information on the chemical composition of matter in space and time. In this respect, remarkable progress has been achieved over the years for the analysis of samples of different kinds based on the condition that they can be brought into the laboratory for analysis with sophisticated analytical techniques. Where depth-resolved information is required, techniques employing electromagnetic radiation are often ideal candidates as in most cases photons can penetrate the sample and their interaction with different layers monitored. For label-free analysis, nuclear magnetic resonance, far-, mid-, or near-infrared spectroscopy as well as Raman scattering are the most important techniques.

Among these, Raman spectroscopy, already in its most simple version of spontaneous Raman scattering, appears to be extremely versatile and applicable to many different scenarios. With the use of confocal Raman microscopy small samples in close proximity to the microscope objective can be analyzed and depth resolutions of typically a few micrometers achieved. Where information on deeper layers is required, other techniques have been developed recently. An instrumentally complex and challenging method for depth resolution is picosecond Kerr gating developed by Matousek et al.,[1] In this case, the sample is probed with a laser and the timing of the 4 ps gate is used to distinguish between the signal of top layers and deeper layers within the sample. This differentiation between the sample container and its content is possible due to the different migration time of the photons in the material, depending on the layer depth. An experimentally more simple approach that also enables recording of Raman spectra from layers several millimeters below the sample surface is spatial offset Raman spectroscopy (SORS), again developed by Matousek et al., in 2005.[2] In their pioneering work, they describe the analysis of trans-stilbene behind a 1 mm layer of PMMA (poly(methyl methacrylate)). The principle of SORS relies on the random scattering of light in a turbid medium. The light propagation depends on the optical density of a material, which influences the probability of a photon to be absorbed or to be converted to a Raman photon at each step. As a consequence, when a laser beam hits a turbid sample, the photons propagate in a random-walk-like fashion.[3]

In Figure 1 a two-layer system is depicted where the excitation laser enters the sample.

Due to random scattering of the photons in the sample material, the illuminated area in the sample increases with increasing depth. In conventional Raman spectroscopy, the detector is placed in line with the laser. However, with SORS the detector is spatially displaced from the laser spot. With increasing offset between laser and detector the ratio of content-to-surface signal rises. This is illustrated in Figure 1 where the percentage of surface signal (black) decreases when the detector is moved from position b to d. At the offset position e, content signal (white) is almost exclusively detected. The theory of this SORS effect has been modeled[4] and experimentally verified. The model shows that an ideal offset position can be located which is strongly dependent on the sample content, container material, and thickness.

The SORS concept can be used for the detection of sugar hidden in diffusely scattering plastic bottles.[5] Furthermore, there are several practical applications, such as transcutaneous in vivo measurement of human bone,[6] identification of counterfeit pharmaceuticals through packaging,[7,8] and detection of concealed ivory.[9] Furthermore, explosives were identified through

Received: August 10, 2011
Accepted: November 3, 2011
Published: November 03, 2011

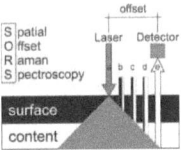

Figure 1. Spatial offset Raman spectroscopy (SORS) principle; when probing a two-layered sample, increasing the distance (offset) between excitation laser and detector favors signal detection from the deeper layer (white) rather than from the surface (black).

a variety of different containers.[10] Even combinations of SERS (surface-enhanced Raman spectroscopy) with SORS were realized for transcutaneous glucose sensing in a rat model.[11] By combining SORS and picosecond gated detection Iping Petterson et al.[12] measured PET (poly(ethylene terephthalate)) through Teflon layers up to 7 mm in thickness. In each of the aforementioned applications there is direct contact of the sample with the measurement apparatus.

The aim of this current work is to extend the application of SORS for investigation of samples at remote distances, 12 m in this paper. Techniques capable of analyzing samples from a distance are advantageous since they reduce the risk for operators as well as for the equipment, when analyzing potentially dangerous samples. Fountain et al.[13] reviews three different stand-off methods for analysis of chemical, biological, and explosive substances on surfaces, namely, LIBS (laser-induced breakdown spectroscopy), fluorescence, and Raman spectroscopy. For a detailed overview on stand-off spectroscopic techniques the reader is referred to a review about Raman spectroscopy by Hobro and Lendl[14] and a review on LIBS by Salle et al.[15] To reduce the limitations of the individual techniques the complementary stand-off methods LIBS and Raman were combined using parametric data fusion for the identification of explosives.[16] Stand-off NIR detection was used for identification of ammonium salts concealed behind layers of clothing.[17] Other stand-off Raman detection methods focused on the detection of analytes either on surfaces,[18] or through glass windows,[19] through transparent plastic bottles,[20] or through colored glass bottles,[21] or during inclement weather conditions.[22] With the use of 10 ps laser pulses it is possible to separate instantaneous Raman signals from fluorescence with a longer lifetime of some nanoseconds.[23] However, the common limitation of stand-off Raman spectroscopy is the necessity of transparent or near-transparent containers. Our experimental system has been shown to detect explosives and their precursors at a distance of 20 m[24] as well as quantification of different analytes in bulk material at remote distances.[25]

With the implementation of SORS, the constraints of transparency-limited, conventional stand-off techniques can be overcome. We present stand-off SORS, a technique capable of analyzing the content within opaque, white, or fluorescent containers at a distance. Plastic containers filled with isopropyl alcohol or NaClO$_3$ were measured at a stand-off distance of 12 m. To further complicate the sampling conditions the plastic container was made highly fluorescent by spreading motor oil on its surface. For the fluorescent sample container it was impossible to record conventional stand-off Raman spectra of the content. However, by increasing the offset it was possible to identify the content despite the challenging interference.

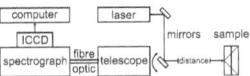

Figure 2. Spatial offset stand-off Raman setup; the laser offset is adjusted with a rotating mirror.

■ **EXPERIMENTAL SECTION**

Chemicals. NaClO$_3$ (>99%) and isopropyl alcohol (>99.5%) were obtained from Sigma-Aldrich, Germany, and a 1 mm extruded polyethylene (PE) board (Wettlinger Kunststoffe, Austria) was used to evaluate the telescope field of view. For sample measurements, an opaque PE container (with a flat surface and wall thickness of 1.5 mm) was filled with isopropyl alcohol, and in the case of NaClO$_3$, a white chemical container made of high-density polyethylene (HDPE) (with a flat surface and wall thickness of 1.5 mm) was used. Motor oil (OMV, Austria) was used as fluorescent sample interference.

Instrumentation. *Pulsed Stand-Off Raman Setup.* The pulsed stand-off Raman system (Figure 2) has been previously described in detail elsewhere,[25] and only a brief description is given here. The excitation laser is a Q-switched Nd:YAG NL301HT laser (EKSPLA, Lithuania) operating at the second harmonic, 532 nm, with a pulse length of 4.4 ns and a repetition rate of 10 Hz.

Raman scattered light was collected using a 6 in. Schmidt−Cassegrain telescope (Celestron, U.S.A.). The light was filtered through a long-pass filter for 532 nm (Semrock, U.S.A.) and collected via a fiber-optical bundle cable consisting of 19 200 μm diameter optical fibers (Avantes, Netherlands). The bundle was directed to an Acton standard series SP-2750 imaging spectrograph. The light was detected by a PIMAX 1024RB intensified charge-coupled device (ICCD) camera (Princeton Instruments, Germany). The camera gate width was set to 5 ns. The laser and ICCD camera were synchronized so that the measurement window coincided with the maximum Raman signal, minimizing the signal contributions from fluorescence and daylight. Spectra were obtained after columnwise pixel binning. The excitation laser intensity was adjusted so as to avoid visible deterioration and destruction of the samples. To ensure fine adjustable laser offset on the sample a mirror was mounted on a rotation stage (Thorlabs, Germany). In this work, the laser pulse energy of 52.6 mJ was focused on an area of 28 mm^2 and 100 pulses were coadded to obtain a spectrum.

■ **RESULTS AND DISCUSSION**

Field of View. For stand-off SORS it is crucial to apply collection optics with a well-defined detection area on the sample, to allow separation of surface and content signal. Therefore, a Schmidt−Cassegrain telescope was used with a limited field of view (FOV). The area of light collection at a distance of 12 m was determined by scanning the laser across a 1 mm thick PE sheet. The maximum collection efficiency is achieved at the zero position (Figure 3), indicating the middle of the telescope.

Increasing offset leads to a reduction of collected Raman light. The baseline-corrected integral of the PE bands in the spectral range from 2500 to 3300 cm^{-1} is shown in Figure 3. By fitting a Gaussian curve through the measurement points, the fwhm (full width at half-maximum) was determined as 11.3 mm. This means

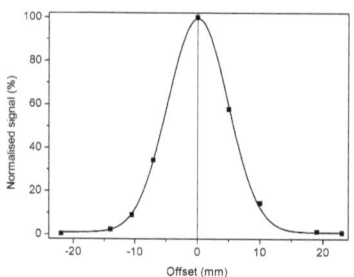

Figure 3. Field of view (FOV) of light collection arrangement; influence of the spatial offset on signal collection efficiency at a stand-off distance of 12 m; baseline-corrected integrals of polyethylene bands between 2500 and 3300 cm^{-1} are drawn as a function of the laser offset relative to the telescope axis. A Gaussian curve was fitted through the measured data, leading to a fwhm (full width at half-maximum) of 11.3 mm or 0.054°.

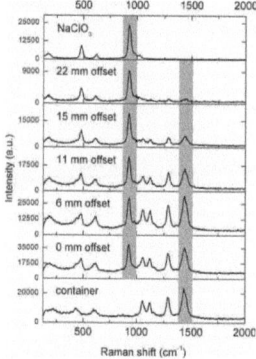

Figure 4. Stand-off SORS spectra of sodium chlorate (NaClO$_3$) in a white HDPE container at 12 m distance; the top and bottom show reference spectra of pure NaClO$_3$ and the container, respectively; other spectra show stand-off SORS spectra; marked bands were used to calculate normalized band intensities.

at an angle 0.027° off the telescope axis on either side, the light collection efficiency is reduced to 50% of the maximum and decreases further to 10% at an offset of 10.6 mm (0.05°).

As a consequence at an offset of 10.6 mm only 10% of the Raman signal generated is collected. However, this is only true for light from the sample surface.

Spatial Offset Raman Scattering. Sodium chlorate (NaClO$_3$) in a white HDPE (high-density polyethylene) bottle was measured at different offsets to demonstrate the SORS effect. In Figure 4 comparison spectra of NaClO$_3$ (top) and the empty container (bottom) are shown together with the stand-off SORS spectra.

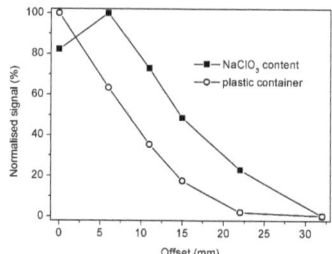

Figure 5. Change of band intensities with offset; signal integrals of the container (1367–1520 cm^{-1}) and NaClO$_3$ (861–998 cm^{-1}) bands; for ease of comparison the maximum integrals were set to 100%.

In Figure 4, at zero offset, the spectral data is a combination of the container and the sample spectral features. Without knowing the spectral signature of the container it would be difficult to identify which bands originate directly from the sample alone. However, it can be observed that with increasing the spatial offset all bands between 1000 and 1500 cm^{-1} (arising primarily from the container) decrease significantly faster than the bands at lower Raman shifts (primarily from NaClO$_3$) making it easier to give a positive identification of the analyte present in the sample. Here the SORS technique has yielded analyte spectral information which would be otherwise difficult or impossible to obtain without prior knowledge of the container spectral data at zero offset alone.

In Figure 5 the influence of the spatial offset on Raman band intensities is shown for the NaClO$_3$ content and the container. Here, baseline-corrected integration of the spectral region between 861 and 998 cm^{-1} was performed to achieve the NaClO$_3$ signal. For better comparability, the maximum NaClO$_3$ integral was set to 100% and the other integrals were normalized accordingly. Similar procedures were performed on the band from the container band (integrated from 1367 to 1520 cm^{-1}) and normalized to the maximum value.

In Figure 5 the signals decay as the offset is increased which would be consistent with limitation of the FOV of the telescope. Here, in this representation of the data and from Figure 4 it can clearly be observed that the container signal decreases more rapidly than that due to the analyte signal consistent with the SORS effect. The signal from the plastic container at an offset of 22 mm is close to zero which is consistent with the FOV determination shown in Figure 3 where practically no light is collected at this offset. NaClO$_3$ signal is still detected at this offset due to the broadening of the excited volume with increasing laser penetration depth. Thus, by obtaining spatially offset spectra the possibility to obtain spectral information of a sample concealed in a container is improved, even at stand-off distances.

Liquid in Container. Isopropyl alcohol in an opaque plastic container was investigated to determine whether there would be a similar effect when the offset is increased. In Figure 6 reference spectra of pure isopropyl alcohol and the empty plastic container are shown (top and bottom spectrum, respectively). The spectrum labeled "0 mm offset" shows the result of a conventional stand-off Raman measurement, where the offset between telescope axis and laser illumination spot is zero.

Analytical Chemistry

ARTICLE

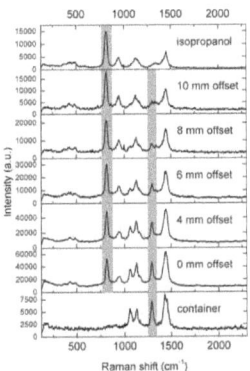

Figure 6. Stand-off SORS spectra of isopropyl alcohol in an opaque plastic container at 12 m distance; the top and bottom show reference spectra of pure isopropyl alcohol and the container, respectively; other spectra show stand-off SORS spectra; marked bands were used to calculate normalized band intensities.

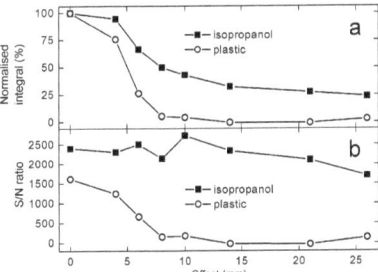

Figure 7. Change of spectral characteristics with offset: (a) normalized signal integrals of isopropyl alcohol (758–881 cm^{-1}) and the plastic container (1248–1341 cm^{-1}); (b) signal-to-noise ratios, rms noise from 1800 to 2200 cm^{-1}.

The data collected with different offsets show the change in relative band intensity. Compared to the isopropyl alcohol band at 816 cm^{-1} the container band at 1291 cm^{-1} decreases quicker with increasing offset. As a result it is possible to distinguish between container and content, even without prior knowledge of the sample. The isopropyl alcohol bands between 758 and 881 cm^{-1} and between 1248 and 1341 cm^{-1} from the plastic container were baseline-corrected and integrated. These signal integrals were normalized to the value at 0 mm offset and are shown in Figure 7a. With increasing offset the band intensity of the isopropyl alcohol decreases slower than the container signal due to the SORS effect.

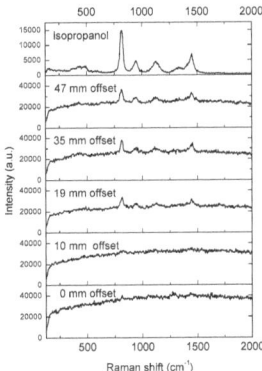

Figure 8. Stand-off SORS spectra of isopropyl alcohol in a highly fluorescent plastic container at 12 m distance: top, reference spectrum of pure isopropyl alcohol; other spectra show stand-off SORS spectra.

Figure 7b shows the signal-to-noise (S/N) ratio of container and content. The root mean square (rms) noise in the spectral range from 1800 to 2200 cm^{-1} was calculated, which decreases with increasing offset. Although the S/N of the container signal decreases the further the laser beam moves out of the telescope axis, the ideal offset for isopropyl alcohol is reflected by the maximum S/N at 10 mm, slowly decreasing at higher offsets. Similar behavior was observed by Maher and Berger[4] who determined ideal offsets for the SORS detection of Teflon behind a layer of dairy creamer. Again, using the SORS principle it is easier to determine the analyte in a concealed opaque container.

Fluorescent Container. In Figure 6 it is still possible to see the signal of the isopropyl alcohol among the container bands when the laser beam is centered with respect to the telescope axis. In order to extend the principle a fluorescent container was selected. In this experiment the container was made highly fluorescent by smearing motor oil on the outside. Figure 8 displays the results for this experiment. As expected at 0 mm offset due to the strong fluorescence the baseline is elevated and no analyte signal, or signal from the container material, can be unambiguously detected. Only at increasing spatial offset between the telescope axis and the sample can the isopropyl alcohol bands be distinguished from the container fluorescent background signal. The detected fluorescence from the sample container decreases by half as the offset changes from 0 to 47 mm. Identification was made possible only by using SORS thus highlighting it as a tool by which disturbing fluorescence from sample containers can be overcome to a certain extent. However, if the source of fluorescence is from the analyte itself this technique would not be so successful.

■ **CONCLUSION**

The feasibility of using SORS at stand-off distances has been demonstrated for the first time. It was shown that the concept of SORS is not limited to close contact measurements. At a distance of 12 m, stand-off SORS spectra of solid as well as liquid material in opaque, white, and fluorescent containers were recorded. This new instrumentation broadens the applicability of SORS for analysis of inaccessible samples from a distance. The content of containers which prevent effective analysis of conventional (zero-offset) stand-off Raman spectroscopy from succeeding can be measured with this novel technique.

■ AUTHOR INFORMATION

Corresponding Author
*Phone: +43 1 58801 15140. Fax: +43 1 58801 15199. E-mail: bernhard.lendl@tuwien.ac.at.

■ ACKNOWLEDGMENT

The research leading to these results has received funding from the European Community's Seventh Framework Program (FP7/2007-2013) under Grant Agreement No. 218037.

■ REFERENCES

(1) Matousek, P.; Everall, N.; Towrie, M.; Parker, W. A. *Appl. Spectrosc.* **2005**, *59*, 200−205.
(2) Matousek, P.; Clark, I. P.; Draper, E. R. C.; Morris, M. D.; Goodship, A. E.; Everald, N.; Towrie, M.; Finney, W. F.; Parker, A. W. *Appl. Spectrosc.* **2005**, *59*, 393−400.
(3) Matousek, P.; Morris, M. D.; Everald, N.; Clark, I. P.; Towrie, M.; Draper, E.; Goodship, A.; Parker, A. W. *Appl. Spectrosc.* **2005**, *59*, 1485−1492.
(4) Maher, J. R.; Berger, A. J. *Appl. Spectrosc.* **2010**, *64*, 61−65.
(5) Eliasson, C.; Macleod, N. A.; Matousek, P. *Vib. Spectrosc.* **2008**, *48*, 8−11.
(6) Matousek, P.; Draper, E. R. C.; Goodship, A. E.; Clark, I. P.; Ronayne, K. L.; Parker, A. W. *Appl. Spectrosc.* **2006**, *60*, 758−763.
(7) Eliasson, C.; Matousek, P. *Anal. Chem.* **2007**, *79*, 1696−1701.
(8) Ricci, C.; Eliasson, C.; Macleod, N. A.; Newton, P. N.; Matousek, P.; Kazarian, S. G. *Anal. Bioanal. Chem.* **2007**, *389*, 1525−1532.
(9) Hargreaves, M. D.; Macleod, N. A.; Brewster, V. L.; Munshi, T.; Edwards, H. G. M.; Matousek, P. *J. Raman Spectrosc.* **2009**, *40*, 1875−1880.
(10) Eliasson, C.; Macleod, N. A.; Matousek, P. *Anal. Chem.* **2007**, *79*, 8185−8189.
(11) Yuen, J. M.; Shah, N. C.; Walsh, J. T.; Glucksberg, M. R.; Van Duyne, R. P. *Anal. Chem.* **2010**, *82*, 267−297.
(12) Iping Petterson, I. E.; Dvořák, P.; Buijs, J. B.; Gooijer, C.; Ariese, F. *Analyst* **2010**, *135*, 3255−3259.
(13) Fountain, A. W., III; Guicheteau, J. A.; Pearman, W. F.; Chyba, T. H.; Christesen, S. D. *Proc. SPIE—Int. Soc. Opt. Eng.* **2010**, *7679*, 76790H−47679H.
(14) Hobro, A. J.; Lendl, B. *Trends Anal. Chem.* **2009**, *28*, 1235−1242.
(15) Salle, B.; Mauchien, P.; Maurice, S. *Spectrochim. Acta, Part B* **2007**, *62*, 739−768.
(16) Moros, J.; Laserna, J. J. *Anal. Chem.* **2011**, *83*, 6275−6285.
(17) Canal, C. M.; Saleem, A.; Green, R. J.; Hutchins, D. A. *Anal. Methods* **2011**, *3*, 84−91.
(18) Pacheco-Londoño, L. C.; Ortiz-Rivera, W.; Primera-Pedrozo, O. M.; Hernández-Rivera, S. P. *Anal. Bioanal. Chem.* **2009**, *395*, 323−335.
(19) Pettersson, A.; Wallin, S.; Östmark, H.; Ehlerding, A.; Johansson, I.; Nordberg, M.; Ellis, H.; Al-Khalili, A. *Proc. SPIE—Int. Soc. Opt. Eng.* **2010**, *7664*, 76641K-1−76641K-12.
(20) Ramirez-Cedeno, M. L.; Ortiz-Rivera, W.; Pacheco-Londono, L. C.; Hernandez-Rivera, S. P. *IEEE Sens. J.* **2010**, *10*, 693−698.
(21) Sharma, S. K.; Misra, A. K.; Sharma, B. *Spectrochim. Acta, Part A* **2005**, *61*, 2404−2412.
(22) Pettersson, A.; Johansson, I.; Wallin, S.; Nordberg, M.; Östmark, H. *Propellants, Explos., Pyrotech.* **2009**, *34*, 297−306.
(23) Akeson, M.; Nordberg, M.; Ehlerding, A.; Nilsson, L. E.; Östmark, H.; Strömbeck, P. *Proc. SPIE—Int. Soc. Opt. Eng.* **2011**, *8017*, 80171C-1−80171C-8.
(24) Zachhuber, B.; Ramer, G.; Hobro, A. J.; Lendl, B. *Proc. SPIE—Int. Soc. Opt. Eng.* **2010**, *7838*, 78380F-1−78380F-10.
(25) Zachhuber, B.; Ramer, G.; Hobro, A. J.; Chrysostom, E. t. H.; Lendl, B. *Anal. Bioanal. Chem.* **2011**, *400*, 2439−2447.

Publication 6

B. Zachhuber, C. Carrillo-Carrión, B.M. Simonet Suau, B. Lendl

Quantification of DNT isomers by capillary liquid chromatography using at-line SERS detection or multivariate analysis of SERS spectra of DNT isomer mixtures

Journal of Raman Spectroscopy, accepted 23/11/2011

Research Article

Received: 11 August 2011 Revised: 23 October 2011 Accepted: 23 November 2011 Published online in Wiley Online Library

(wileyonlinelibrary.com) DOI 10.1002/jrs.3149

Quantification of DNT isomers by capillary liquid chromatography using at-line SERS detection or multivariate analysis of SERS spectra of DNT isomer mixtures

Bernhard Zachhuber,[a] Carolina Carrillo-Carrión,[b]* Bartolomé M. Simonet Suau[b] and Bernhard Lendl[a]*

With the global surge of terrorism and the increased use of bombs in terrorist attacks, national defence and security departments now demand techniques for quick and reliable analysis, in particular, for detection of toxic and explosive substances. One approach is to separate different analytes and matrix material before detection. In this work microliquid chromatography was used to separate two dinitrotoluene (DNT) isomers prior to detection via online UV–Vis spectroscopy. For identification, retention times were compared with reference samples and quantification was done by integration of UV–Vis absorption. Because UV detection is not particularly selective, Raman microscopic analysis was coupled to the liquid chromatography using a flow-through microdispenser. Because DNT is difficult to detect with conventional Raman spectroscopy, the sensitivity was increased via surface-enhanced Raman scattering (SERS) using silver-quantum dots. Different analytical approaches to identify and quantify mixtures of two DNT isomers were evaluated. Good quantitative results were obtained using UV detection after microchromatographic separation (Limit of Detection: 0.11 and 0.06 for 2,4-DNT and 2,6-DNT). Coupling with SERS allowed for more confident differentiation between the highly structurally similar DNT isomers because of the additional spectral information provided by SERS. The application of a partial least squares algorithm also allowed direct SERS detection of DNT mixtures (root mean square error of prediction: 0.82 and 0.79 mg·L^{-1} for 2,4-DNT and 2,6-DNT), circumventing the time-consuming separation step completely. Copyright © 2011 John Wiley & Sons, Ltd.

Keywords: silver-quantum dot colloid (Ag-QD); SERS detection; dinitrotoluene (DNT); PLS multivariate regression

Introduction

With the increase of terrorist attacks, the development of new, cost-efficient devices for rapid explosive detection has become an urgent necessity.[1] Various physical methods, such as gas chromatography coupled with a mass spectrometer, nuclear quadrupole resonance, energy-dispersive X-ray diffraction and electron capture detection have been used for this purpose.[2-5] The most common method for the determination of explosives is reverse phase high-performance liquid chromatography (HPLC) with UV detection.[6] Comparing retention times with references provides evidence for the chemical identity, whereas quantification is achieved via UV–Vis absorption at appropriate wavelengths. Because of the nondestructive character of liquid chromatography (LC), spectroscopic techniques such as Fourier transform infrared and Raman spectroscopy can be added to the existing separation system to gain detailed structural information, for example to distinguish isomers. Here, 2,4-DNT and 2,6-DNT (dinitrotoluene) were investigated because they are degradation products and impurities in TNT (trinitrotoluene) and are used for the production of landmines. Unfortunately, DNT and similar classes of compounds are usually not detected by conventional Raman measurements[7] but this lack of sensitivity can at least partly be overcome using surface-enhanced Raman scattering (SERS), which can also improve the selectivity.[8] Raman signals of molecules within highly localized optical fields of metallic nanostructures can be enhanced by factors of 10^3 to 10^6 because of the effects of electromagnetic field and chemical enhancement.[9-11] Moreover, with SERS fluorescence quenching and reduced interference of solvent bands can lead to limits of detection (LOD) on the ppb level or even single molecule level.[12] A recently reported silver-quantum dot colloid[13] was used for the detection of pesticides and nucleic bases. The sponge morphology of this SERS substrate led to significant SERS enhancement, associated with the resonance coupling between neighboring nanoparticles localized on the pores (hot spots). When coupled with infrared (IR) and Raman detection the trend towards miniaturized separation systems such as capillary and nano-HPLC systems increases the sensitivity because of small dead volumes. Furthermore, the required sample amount is

---

* Correspondence to: Bernhard Lendl, Institute of Chemical Technologies and Analytics, Vienna University of Technology, A-1060 Vienna, Austria. E-mail: bernhard.lendl@tuwien.ac.at

Carolina Carrillo-Carrión, Department of Analytical Chemistry, University of Córdoba, E-14071 Córdoba, Spain. E-mail: q12cacac@uco.es

a Institute of Chemical Technologies and Analytics, Vienna University of Technology, A-1060 Vienna, Austria

b Department of Analytical Chemistry, University of Córdoba, E-14071 Córdoba, Spain

drastically reduced for capillary-LC because of the low flow rates of about 3 μL·min⁻¹. For the reproducible placement of the eluting analyte from the chromatographic column on the SERS substrates, a flow-through microdispenser was used in this work. A concept that has already been described for the coupling of micro-LC and Raman, and IR detection.[14] Here, μ-LC was used for the separation of 2,4-DNT and 2,6-DNT followed by the quantification and identification using either UV absorption or SERS detection using Ag-QD SERS.[13] Furthermore, SERS spectra of physically unseparated DNT mixtures were recorded. Employing a partial least squares (PLS) regression the chromatographic separation was circumvented.

Experimental

Reagents and materials

All chemical reagents were of analytical grade and used as purchased with no additional purification. Cadmium oxide (99.99%), trioctylphosphine oxide (TOPO, 99%), trioctylphosphine (TOP, 90%), selenium (powder, 100 mesh, 99.99%), diethylzinc solution (ZnEt$_2$, ~1 M in hexane), bis(trimethylsilyl) sulphide ((TMS)$_2$S), anhydrous methanol and anhydrous chloroform were purchased from Sigma Aldrich (Madrid, Spain). Hexylphosphonic acid was obtained from Alfa Aesar (Karlsruhe, Germany). Methanol (HPLC grade), silver nitrate (99.5%) hydroxylamine hydrochloride (98%) 2,4-DNT (97%) and 2,6-DNT (99.4%) were purchased from Sigma Aldrich (Schnelldorf, Germany).

Stock solutions of the two DNT isomers (1000 mg·L⁻¹) were prepared in acetonitrile and stored at 4 °C. Diluted standards were prepared from these solutions by appropriate dilution in distilled water.

The SERS colloid used consisted of a silver-quantum dots solution (Ag-QD), which has already been demonstrated to be a very active and reproducible substrate for SERS measurements. Briefly, it was prepared by reducing of Ag(I) to Ag(0) in the presence of ZnS-capped CdSe QDs previously dispersed with hydroxylamine as stabilizing agent. Then, the SERS-active substrate was prepared by filling 20 μL of the Ag-QD colloidal solution into each well of a home-made microtiter plate. The microplate consisted of a teflon plate with conical wells in a line. The size of each well was 3 mm in diameter and 3 mm in depth and the distance between wells was 0.5 mm.

Instrumentation

The μLC-UV chromatographic system consisted of an Ultimate 3000 Dionex with a 1 μL injection loop and a C18 Acclaim PepMap (300 μm ID × 15 cm, 3 μm, 100 Å) separation column from Dionex. The column was kept at 25 °C and the experiments were performed with 3 μL·min⁻¹ eluent flow rate. For the separation, methanol was used as solvent A and deionized water was solvent B and the following gradient program were used: 0 min 50% A, 15 min 80% A, 20 min 80% A, 22 min 50% A and 30 min 50% A. All the eluents were transferred to the microdispenser without flow splitting. The UV–Vis detector (Ultimate UV detector, LC Packings, Dionex) was set at 245 nm and placed just before the microdispenser interface. An in-house developed program based on LabVIEW 8.5 software (National Instruments, Austin, USA) was used to control the UV spectrometer and register the chromatograms.

The flow-through microdispenser interface is a microliquid handling device formed by two conventionally micromachined silicon structures. The dispenser was driven by a dc power supply (HGL 5630 DLBN) together with a computer-controlled arbitrary waveform generator (Agilent 33120A, Agilent Technologies, Palo Alto, CA), which provided an electronic pulse with defined amplitude (15 V). A computer-controlled x,y-stage (Newport THK, Compact Linear Axis) with step sizes of 5 μm and a maximum distance range of 90 mm × 40 mm was implemented in the dispensing unit. The connection between the microdispenser and the μLC column was via a fused silica capillary (i.d. 50 μm, o.d. 364 μm, and 40 cm long). All the computer-controlled components of the microdispensing unit were operated via an in-house-written MS Visual Basic 6.0 (Microsoft) based software program (SAGITTARIUS, Version 3.0.25) working under Windows NT. The optimal working conditions for the microdispenser-stage unit were as follows: it waited for 950 s before a quick movement of 3 mm at 2000 μm per second followed by a stop of 40 s with a dispensing frequency of 100 droplets per second, this cycle was repeated five additional times.

Raman measurements were acquired with a confocal Raman microscope (LabRaman, Horiba Jobin Yvon Ltd., Bensheim, Germany) using a 633 nm laser line and a charge-coupled device detector with 1024×256 pixels. A grating with 600 grooves/mm, a confocal aperture of 500 μm and an entrance slit of 100 μm were selected for the experiments. A 10× microscope objective was used to focus the laser beam (20 mW) on each well of the microtiter plate. In all cases spectra were recorded with a data acquisition time of 20 s and were baseline corrected.

Results and Discussion

μLC separation and UV detection (μLC-UV)

Mobile phases of water–acetonitrile and water–methanol modified with 2-propanol or formic acid were tested for separation of 2,4-DNT and 2,6-DNT. For each mobile phase, the modifier proportion and the eluent program were varied to find optimal separation conditions. For many conditions tested the two analytes coeluted or showed poor resolution. The best resolution (Rs = 0.8) was achieved with mobile phase water–methanol and the following gradient program: 50% methanol at 0 min, 80% methanol at 15 min, 80% methanol at 20 min, 50% methanol at 22 min and 50% methanol at 30 min. The flow rate was 3 μL·min⁻¹ and UV detection was performed at 245 nm. The resolution (Rs) was calculated as: $Rs = 2(t_2 - t_1)/(w_2 + w_1)$, where t_2 and t_1 are the retention times of the isomers and w_2 and w_1 are the corresponding peak widths at the baseline. Figure 1(A) shows a typical chromatogram of the two isomers recorded by absorbance detection at 245 nm when a standard solution of 30 mg·L⁻¹ (30 ng injected in column) for each analyte was injected.

The analytical features of the μLC-UV method in terms of sensibility, LOD and quantification (LOQ) and precision were studied and the data are presented in Table 1. Calibration graphs were constructed by plotting the peak areas of each isomer, measured at 245 nm, versus the concentration injected in column. The LOD and LOQ were calculated from the standard deviation of seven measurements of the peak area at a concentration level of 20 mg·L⁻¹ divided by the sensitivity and multiplying by 3 or 10 for LOD and LOQ, respectively.

Quantification of DNT isomers by capillary liquid chromatography using at-line SERS detection

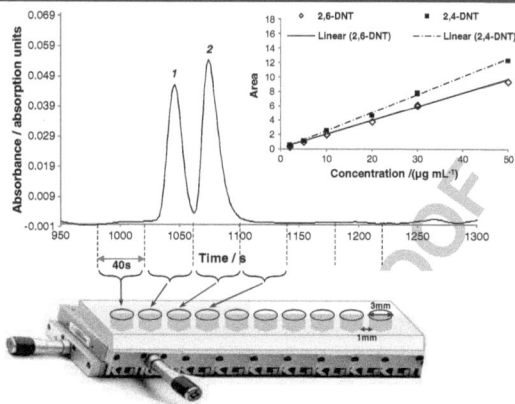

Figure 1. Chromatogram obtained after injecting a mixture of standard solution of the two DNT isomers at 30 mg·L^{-1}. Peaks: 1 = 2,6-DNT (t_r = 17.4 min); 2 = 2,4-DNT (t_r = 17.9 min). Conditions: water–methanol (gradient program); 3 µL·min^{-1}. Detection wavelength: 245 nm.

Table 1. Comparison of evaluation parameters of the three methodologies studied: (1) µLC-UV, (2) µLC-(microdispenser)-SERS and (3) SERS-multivariate PLS for the determination of the two DNT isomers

	µLC-UV		µLC-(microdispenser)-SERS	
	2,6-DNT	2,4-DNT	2,6-DNT	2,4-DNT
Retention time (min)	17.4	17.9	17.4	17.9
Analytical signal	Absorbance (245 nm)	Absorbance (245 nm)	Raman Int. (2950 cm^{-1})	Raman Int. (2950 cm^{-1})
Regression equation	$y = 0.192x + 0.096$	$y = 0.251x + 0.015$	$y = 761.5x + 0.2$	$y = 488.0x + 0.7$
Linearity (R^2)	0.997	0.997	0.996	0.988
LOD (mg·L^{-1})	0.11	0.06	0.51	0.54
LOQ (mg·L^{-1})	0.37	0.20	1.70	1.82
RSD (%)	2.8	1.6	3.7	3.8

µLC separation and SERS detection (µLC-microdispenser-SERS)

Microdispenser–microplate collection

The chromatographic separation was transferred into a microplate via a flow-through microdispenser. With the help of a fully automated xy-stage, the eluent from the column was collected in the wells of a home-made microtiter plate. The working conditions of the microdispenser stage were adjusted to collect the total amount of each analyte in separate, individual wells. A schematic picture of the fraction collected in each well is presented under the chromatogram in Figure 1.

Surface-enhanced Raman scattering measurements

To prepare the SERS-substrate 20 µL of the Ag-QD solution were placed into each well of the microplate before collecting the analytes from the µLC-microdispenser. With an injection volume of 1 µL, without flow column splitting and using a high dispersing frequency of the microdispenser, it was considered that the total amount of each analyte injected in the µL column was collected in the microplate.

The Raman spectra of the studied analytes in solution could only be recorded using the silver-quantum dot colloidal solution. Without the enhancement associated with the nanoparticles, Raman signals in solution were not obtained. Alternatively, when using silver colloidal solution (prepared by Leopold-Lendl method) SERS signal could only be observed at high analyte concentration and signal was still weak. Therefore, the large Raman enhancement of the Ag-QD solution was necessary to obtain good signals for SERS detection of the two DNT isomers in solution.

The methodology µLC-(microdispenser)-SERS was also evaluated in terms of sensitivity, LOD, LOQ and precision and the results obtained are summarized in Table 1. For these evaluations, different concentrations were used and five independent solutions of 40 mg·L^{-1} were analysed using the whole µLC-(microdispenser)-SERS procedure. The analytical signal used was the Raman intensity at 2950 cm^{-1}. Calibration graphs were constructed by plotting the Raman intensity of the selected peak versus the concentration injected in the µLC.

SERS detection and PLS data treatment (SERS-multivariate PLS)

Whereas the two previous methods necessitate analyte separation via μLC, the direct detection with SERS circumvents the time-consuming preparation step, thereby increasing the sample throughput.

The weak DNT Raman signal was amplified by SERS, leading to spectral data quality sufficient for quantification. The spectral profile of both isomers was relatively similar (Fig. 2) necessitating further data treatment to separate 2,4 DNT and 2,6 DNT signals. For this multivariate calibration, a PLS algorithm (OPUS 6.5; Bruker, Germany; Quant 2 method) was employed. The calculated model was based on a training set of 34 different isomer mixtures, ranging from 0 to 40 mg·L^{-1}. The following spectral regions of the training set were incorporated into the model without prior data processing, avoiding bands of acetonitrile, which was used to dissolve the DNT isomers: 400–890, 950–1300 and 1500–1700 cm^{-1}. To validate the model, 20 different DNT mixtures were prepared. The determination coefficient (R^2), the root mean square error of estimation (RMSEE) the root mean square error of prediction (RMSEP) and the residual prediction deviation (RPD) were calculated to determine the quality of the multivariate calibration.

The summarised results of the multivariate calibration of 2,4 DNT and 2,6 DNT in Table 2 show that the RPD are in the range of 9%. Furthermore, the RMSEP shows that the model predicts the concentration of either isomer in an unseparated mixture with an average accuracy better than 1 mg·L^{-1}.

Comparison

When comparing the three methods for detection of 2,4-DNT and 2,6-DNT the best figures of merit are achieved with online UV detection. Compared with SERS detection after μLC separation, online UV detection leads to a LOD and a LOQ nine times better for 2,4-DNT and 4.6 times better for 2,6-DNT. The most important benefit of SERS detection lies in the spectral information gained. Whereas UV detection at specific wavelength requires previous knowledge of the substances, SERS detection adds structural information. Therefore, in addition to quantitative information, identification of analytes can be achieved. Because of the spectral features measured via SERS multivariate analysis is also possible. This means that no μLC separation is needed prior to detection. The isomers are analysed simultaneously and the spectral information is separated mathematically. The established multivariate model leads to RPD of 8.9% for 2,4-DNT and 9.5% for 2,6-DNT. The higher error is compensated by a much greater sample throughput because of the circumvention of the relatively time-consuming μLC separation.

If a high sample throughput is not the primary aim the combination of μLC with online UV detection and additional SERS detection leads to good quantitative results and to additional structural information. Depending on the specific situation one can choose which of the three methods discussed meets the specific requirements best.

Conclusion

The potential of combining different analytical approaches, i.e. μLC-UV, μLC-SERS and SERS-PLS, for the analysis of DNT isomers has been demonstrated. These combinations of techniques provided good quantitative results and additional structural information.

Two especially important aspects in this work were the use of a flow-through microdispenser as interface between a μLC and a Raman microscope, and the type of SERS substrate employed for generating SERS spectra. Without such robustness, SERS detection will not gain widespread acceptance in a combined system for analytical chemistry. Although the μLC-SERS methodology did not achieve lower detection limits compared with μLC-UV, the SERS spectra provided additional structural information. Therefore, the μLC-SERS approach is a complementary detection technique to μLC-UV to confirm the identification of studied analytes and to differentiate between isomers, which tend to co-elute in chromatographic separations.

In addition, because of the high sensitivity and reproducibility achieved in SERS measurements with the 'Ag-QDs' colloid, it was possible to circumvent the chromatographic separation step and resolve mixtures of the two isomers by employing PLS regression, which resulted in a significant reduction of measurement time required.

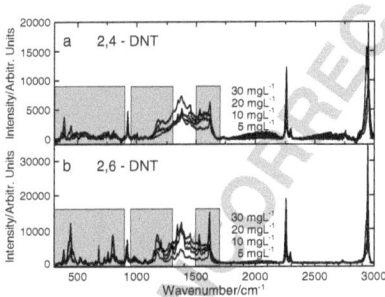

Figure 2. SERS spectra of (a) 2,4-DNT and (b) 2,6-DNT ranging from 5 to 30 mg/l; rectangles illustrate spectral ranges used for multivariate analysis.

Table 2. Validation of multivariate PLS model; determination coefficient (R^2), the RMSEE, RMSEP and the RPD are listed for the training set and the validation set

	Training set		Validation set	
	2,6-DNT	2,4-DNT	2,6-DNT	2,4-DNT
Analytical signal	(400–890) + (950–1300) + (1500–1700) cm^{-1}			
Determination coefficient R^2	98.9	98.75	99.01	98.99
RMSEE (mg·L^{-1})	1.46	0.734	—	—
RMSEP (mg·L^{-1})	—	—	0.82	0.79
RPD (%)	9.5	8.9	—	—

Acknowledgements

C. Carrillo-Carrión would like to express gratitude to the Spanish Ministry of Innovation and Science for the award of a Research Training Fellowship (Grant AP2006-02351). This research was also supported in part by OPTIX, which is funded by the European Community's Seventh Framework; programme (FP7/2007-2013) under grant agreement n° [218037]

References

[1] J. Yinon, *Anal. Chem.* **2003**, *75*, 99A.
[2] K. Hakansson, R. V. Coorey, R. A. Zubarev, V. L. Talrose, P. J. Hakansson, *Mass Spectrom.* **2000**, *35*, 337.
[3] V. P. Anferov, G. V. Mozjoukhine, R. Fisher, *Rev. Sci. Instrum.* **2000**, *71*, 1656.
[4] R. D. Luggar, M. J. Farquharson, J. A. Horrocks, R. J. Lacey, *J. X-ray Spectrom.* **1998**, *27*, 87.
[5] A. M. Rouhi, *Chem. Eng. News* **1997**, *75*, 14.
[6] D. R. Felt, S. L. Larson, L. Escalon, *Talanta* **2008**, *76*, 21.
[7] D. S. Moore, *Rev. Sci. Instrum.* **2004**, *75*, 2499.
[8] B. Sägmüller, B. Schwarze, G. Brehm, G. Trachta, S. Schneider, *J. Mol. Struct.* **2003**, *279*, 661.
[9] C. L. Haynes, C. R. Yonzon, X. Zhang, R. P. Van Duyne, *J. Raman Spectrosc.* **2005**, *36*, 471.
[10] H. Xu, J. Aizpurua, M. Käll, P. Apell, *Physical Review E* **2000**, *62*, 4318.
[11] W. E. Doering, S. Nie, *J. Phys. Chem. B* **2002**, *106*, 311.
[12] K. Kneipp, Y. Wang, H. Kneipp, L. T. Perelman, I. Itzkan, R. R. Dasari, M. S. Feld, *Phys. Rev. Lett.* **1997**, *78*, 1667.
[13] C. C. Carillo-Carrión, B. M. Simonet, M. Valcárcel, B. Lendl, *Anal. Chem.* **2011**, DOI: 10.1021/ac201134d.
[14] I. Surowiec, J. R. Baena, J. Frank, T. Laurell, J. Nilsson, M. Trojanowicz, B. Lendl, *J. Chromatogr. A* **2005**, *1080*, 132.

Publication 7

B. Zachhuber, C. Gasser, E. t. H. Chrysostom, B. Lendl

Spatial Offset Stand Off Raman Scattering

in Lasers, Sources, and Related Photonic Devices, OSA Technical Digest (CD) (Optical Society of America, 2012), paper LT2B.4

Stand-off Spatial Offset Raman Scattering

Bernhard Zachhuber, Christoph Gasser, Engelene t. H. Chrysostom, Bernhard Lendl
Vienna University of Technology, Getreidemarkt 9/164AC, A-1060 Vienna, Austria; +43 1 58801 15140
Bernhard.Lendl@tuwien.ac.at

Abstract: Identification and quantification of potentially harmful substances concealed in fluorescent containers were achieved at a distance of 12 metres using stand-off spatial offset Raman scattering.
OCIS codes: (280.1545) Remote sensing and sensors, Chemical analysis; (300.6450) Spectroscopy, Raman

1. Introduction

We present a method for the stand-off detection (at a distance of 12 meters) of potentially hazardous materials concealed in containers or hidden behind opaque layers using a combination of stand-off Raman spectroscopy [1][2] and Spatial Offset Raman Spectroscopy (SORS) [3]. This combination makes it possible to detect chemicals concealed in containers from a distance where a potential danger to the operator is reduced. SORS, developed by Matousek et al. [4] enables the detection of deeper layers of a sample by placing the collection optics at a position different (spatially offset) to the excitation spot on the sample surface. In our experimental system, stand-off Raman spectra at distances up to 12 metres at different spatial offsets (spatial offsets up to 40 mm away from the incident laser excitation) were collected using a telescope. Merging SORS with stand-off Raman spectroscopy, it was possible to detect liquid (isopropanol, hydrogen peroxide) as well as solid analytes (sodium chlorate) in opaque or white containers. This technique is not limited to qualitative analysis alone but can also be used for the quantification of hydrogen peroxide in an opaque container. Furthermore, the content of fluorescent containers, which is difficult, if not impossible for conventional (zero-offset) stand-off Raman spectroscopy measurements, can now be detected as demonstrated using isopropanol concealed in a plastic bottle covered with a highly fluorescent layer of motor oil.

1. Experimental Setup

The pulsed stand-off Raman system (Fig. 1a) has been previously described in detail elsewhere [5] and only a brief description is given here. The excitation laser is a Q-switched Nd:YAG NL301HT laser (EKSPLA, Lithuania) operating at the second harmonic, 532nm, with a pulse length of 4.4 ns and a repetition rate of 10 Hz. Raman scattered light was collected using a 6″ Schmidt-Cassegrain telescope (Celestron, USA). The light was filtered through a long pass filter for 532 nm (Semrock, USA) and collected via a fibre optical bundle cable consisting of 19 200 μm-diameter optical fibres (Avantes, Netherlands). The bundle was directed to an Acton standard series SP-2750 imaging spectrograph. The light was detected by a PIMAX 1024RB intensified CCD (ICCD) camera (Princeton Instruments, Germany). The camera gate width was set to 5 ns. The laser and ICCD camera were synchronised so that the measurement window coincided with the maximum Raman signal, minimising the signal contributions from fluorescence and daylight. Spectra were obtained after column wise pixel binning. The excitation laser intensity was adjusted so as to avoid visible deterioration and destruction of the samples. To assure fine adjustable laser offset on the sample a mirror was mounted on a computer controlled, motorised mirror mount (Thorlabs, Germany). In this work, the laser pulse energy was set to 52.6 mJ and 100 pulses were co-added to obtain a spectrum.

The SORS principle is illustrated in Fig. 1b. Incident laser beam photons hitting a turbid container wall are strongly scattered, leading to a spatial spread of the laser beam in the probed material. Whereas in conventional Raman spectroscopy the inelastically scattered Raman signal is collected at the point (1) where the laser enters the material, with SORS, the detector is spatially moved away from this point (2). As a consequence the light reaching the detector originates primarily from deeper sample layers when laser spot and detector are spatially offset. The spectra depicted in Fig. 1c illustrate the change in relative band intensities when probing isopropanol in a turbid polyethylene (PE) bottle. At 0 mm offset (1) the Raman bands (highlighted with rectangles) of container (PE) and isopropanol content are both present. But as soon as laser and telescope are 14 mm offset (2) no PE signal is left, whereas the content signal is still clearly visible.

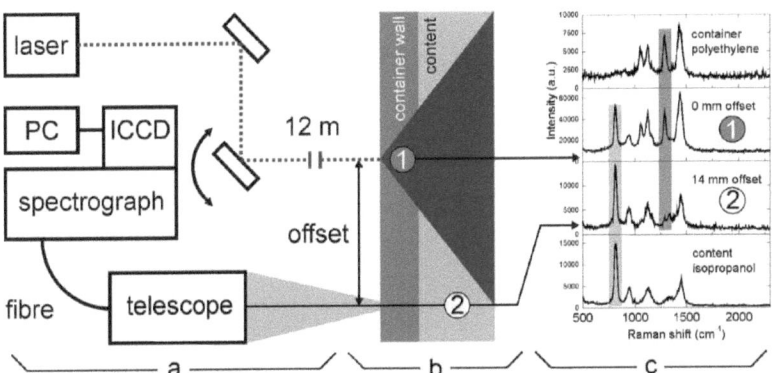

Fig. 1: Stand-off spatial offset Raman spectroscopy; a: setup, b: SORS principle; c: example spectra of isopropanol in a polyethylene container at 12 metres with different spatial offset

3. Results and Discussion

Stand-off SORS spectra of the clear liquid isopropanol are shown in Fig. 1c above. The increase of spatial offset between laser excitation and light collection changes the relative intensities of bottle and content, allowing a clearer detection of concealed substances when the container gives obstructing Raman or fluorescence signals. In Fig. 2 solid sodium chlorate ($NaClO_3$) was placed in a white polyethylene bottle with a wall thickness of 1.5 mm.

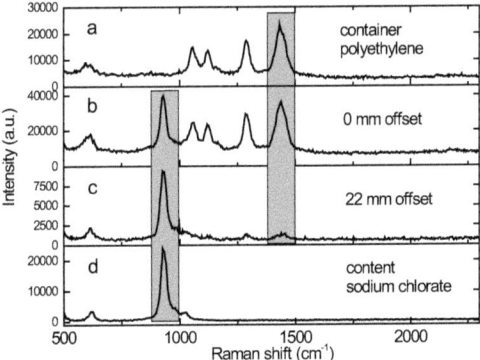

Fig. 2: 12 m-Raman stand-off spectra; a: empty polyethylene (PE) bottle; b: 0 offset spectrum of PE bottle filled with $NaClO_3$ (sodium chlorate); c: SORS spectrum of $NaClO_3$ in PE; d: spectrum of pure $NaClO_3$

In Fig. 2 significant differences in the relative band intensities of content ($NaClO_3$) and container (white PE) can be observed at a spatial offset of 0 and 22 mm respectively. The spectrum obtained at 0 mm offset (Fig. 2b) when compared with the one at the higher offset in Fig. 2c, the sodium chlorate bands are much more pronounced than those bands due to the container thus enabling clear identification of the content ($NaClO_3$).

The potential of stand-off SORS (at 12 m) can also be demonstrated by smearing highly fluorescent motor oil on the outside of a PE bottle containing H_2O_2. At zero offset, the Raman signal was dominated by the fluorescence from the container. By increasing the spatial offset it was possible to avoid the disturbing fluorescent signal from the container wall and successfully detect H_2O_2. A quantitative analysis of H_2O_2 (0-30%) in this fluorescent container was possible only at a spatial offset of 40 mm.

4. Conclusion

In this work the combination of stand-off Raman spectroscopy and Spatial Offset Raman Spectroscopy (SORS) was introduced as a novel method combination for the detection of concealed content in containers at stand-off distances of 12 m. Both solid ($NaClO_3$) and liquid (isopropanol and hydrogen peroxide) samples were detected in opaque sample containers. In addition, a quantitative analysis of hydrogen peroxide in fluorescent container was also performed.

5. References

[1] A. Pettersson, S. Wallin, H. Östmark, A. Ehlerding, I. Johansson, M. Nordberg, H. Ellis, A. Al-Khalili, "Explosives standoff detection using Raman spectroscopy: from bulk towards trace detection" in *Proc. of SPIE* **7664**, 76641K-1-76641K-12 (2010).

[2] B. Zachhuber, G. Ramer, A. J. Hobro, and B. Lendl, "Stand-off Raman spectroscopy of explosives" in *Proc. of SPIE* **7838**, 78380F-1-78380F-10 (2010).

[3] Matousek, M. D. Morris, N. Everall, I. P. Clark, M. Towrie, E. Draper, A. Goodship, A. W. Parker, "Numerical simulations of subsurface probing in diffusely scattering media using spatially offset Raman spectroscopy" Applied spectroscopy **59**, 1485-1492 (2005).

[4] P. Matousek, I. P. Clark, E. R. C. Draper, M. D. Morris, A. E. Goodship, N. Everall, M. Towrie, W. F. Finney, A. W. Parker, "Subsurface Probing in Diffusely Scattering Media Using Spatially Offset Raman Spectroscopy," Applied Spectroscopy **59**, 393-400 (2005).

[5] B. Zachhuber, G. Ramer, A. Hobro, E. t. H. Chrysostom, B. Lendl, "Stand-off Raman spectroscopy: a powerful technique for qualitative and quantitative analysis of inorganic and organic compounds including explosives" Analytical and bioanalytical chemistry **400**, 2439-2447 (2011).

Publication 8

B. *Zachhuber*, C. Gasser, G. Ramer, E. t. H. Chrysostom and B. Lendl

Depth profiling for the identification of unknown substances and concealed content at remote distances using time resolved stand-off Raman spectroscopy

Applied Spectroscopy, accepted 19/04/2012

Depth profiling for the identification of unknown substances and concealed content at remote distances using time-resolved stand-off Raman spectroscopy

Bernhard Zachhuber, Christoph Gasser, Georg Ramer, Engelene t. H. Chrysostom and Bernhard Lendl*

Vienna University of Technology, Institute of Chemical Technologies and Analytics, Getreidemarkt 9/164AC, A-1060 Vienna, Austria; *bernhard.lendl@tuwien.ac.at

Corresponding Author: bernhard.lendl@tuwien.ac.at
Telephone: +43 (0) 15880115140
Fax: +43 (0) 15880115199

Abstract

Time-resolved stand-off Raman spectroscopy was used to determine both the position and identity of substances relative to each other at remote distances (up to tens of metres). Spectral information of three xylene isomers, toluene and sodium chlorate was obtained at a distance of 12 m from the setup. Pairs and triplets of these samples were placed at varying distances (10-60 cm) relative to each other. Via the photon time of flight the distance between the individual samples was determined to an accuracy of 7% of the physically measured distance. Furthermore, at a distance of 40 m, time-resolved Raman depth profiling was used to detect sodium chlorate in a white plastic container which was non-transparent to the human eye. The combination of the ranging capabilities of Raman LIDAR (sample location usually determined using prior knowledge of the analyte of interest) with stand-off Raman spectroscopy (analyte detection at remote distances) provides the capability for depth profile identification of unknown substances and analysis of concealed content in distant objects. To achieve these results, a 532 nm-laser with a pulse length of 4.4 ns was synchronised to an intensified charge coupled device camera with a minimum gate width of 500 ps. For automated data analysis a multivariate curve resolution algorithm was employed.

Keywords: stand-off, remote, Raman LIDAR

Introduction

Optical methods (such as infrared and Raman spectroscopy) have seen increasing application in analytical chemistry and process analytical chemistry applications. One major advantage of these optical methods is that direct sample contact is not necessary. This marks an important difference to well established laboratory based analysis methods which in many cases use mass spectrometric detection schemes. In fact several optical techniques have been used to determine sample positions up to distances of several kilometers such as LIDAR (light detection and ranging) or LADAR (laser detection and ranging). A scanning laser allows the calculation of 3D models of the environment via the photon time of flight.[1] This principle can even be applied under water to find sea mines.[2] These techniques determine the geometric shape of the investigated objects. Other methods allow substance specific ranging such as DIAL (differential absorption lidar) where it is possible to obtain profiles of gas concentration in the atmosphere. Basically two pulsed lasers are used alternatingly. One laser is designed to be absorbed by the molecule of interest (on-line wavelength) the other laser wavelength is chosen to exhibit as little absorption as possible (off-line wavelength), acting as a reference. DIAL allows for example the determination of ozone profiles in the atmosphere.[3] DIAL makes use of molecular specific absorption of light, however for this technique the laser needs to be tuned to the specific wavelength of the analyte as only the elastic part of the back scattered radiation is detected. In comparison, using Raman spectroscopy, samples are excited with one laser wavelength but the inelastically scattered polychromatic light containing molecular specific information is detected. In Raman LIDAR, several substances can be detected using a dedicated PMT (photomultiplier tube) for each substance of interest, i.e. H_2O, N_2 and O_2.[4] PMTs are very sensitive detectors and must be placed at the substance specific wavelength. This necessitates prior knowledge of the substances to be detected.

Via stand-off Raman spectroscopy, the complete spectrum is obtained which permits the identification of unknown substances. UV excitation with CCD detectors was successful in the detection of chemicals both in bulk quantities such as Teflon, cyclohexane and thin contaminant films (acetonitrile on Teflon several microns thick) at stand-off distances ranging from 3 to 30 m.[5,6] Carbonates, silicates, hydrous silicates, and sulfate minerals were detected at stand-off distances ranging from 10 to 66 m using 532nm as the excitation wavelength.[7] Explosives such as RDX, TNT and PETN were detected at 27 and 50 m respectively.[8] Stand-off Raman spectroscopy was also used to determine the content in transparent glass bottles i.e. benzene at 10 m and naphthalene at 100 m.[9,10] Even under challenging weather conditions including fog and rain at remote distances up to 55 m explosives could be identified.[11] Others have shown the advantage of using short laser pulses[12] and narrow ICCD detector gates for stand-off Raman detection of highly luminescent explosives.[13] In addition, the advantageous combination of Raman spectroscopy and LIBS (Laser Induced Breakdown Spectroscopy) for the detection of explosives has been demonstrated.[14,15]

The theory and techniques of time-resolved spectroscopy have been extensively discussed before.[16] This paper however focuses on the use of time-resolved stand-off Raman spectroscopy to determine the position as well as the chemical composition of substances without any previous knowledge of their chemical nature. This is made possible since the complete Raman spectrum is recorded. This is possible since a CCD array is used to collect all inelastically scattered spectral lines instead of only a few selected bands in Raman LIDAR. At close proximity, time-resolved Raman spectroscopy was used to investigate multi-layer systems.[17,18] Here this concept is extended to a remote distance of 40 m. The principle of time-resolved stand-off Raman is illustrated in Fig. 1 where two samples are placed at different distances from the detection system.

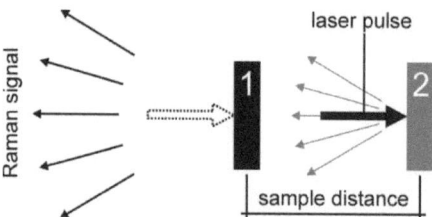

Fig. 1: Time-resolved Raman spectroscopy principle; sample 1 is placed 12 m from the measurement setup

A laser pulse excites a substance thus yielding a specific Raman signal in sample 1. While the same laser pulse travels on to sample 2, the Raman signal of sample 1 already returns to the detection system. The introduced time delay between the two Raman signals permits calculation of the distance between the samples, using the speed of light. For a suitable determination of the sample position it is essential to have a fast gated detection system, synchronised to the pulsed laser which effectively reduces and in some cases rejects completely the contribution due to fluorescence and ambient light. Time-resolved stand-off Raman spectroscopy provides chemical information and depth profile analysis at stand-off distances (at tens of metres). This means that at remote distances the chemical composition as well as the position of unknown objects can be determined. In addition the detection of concealed content in remote objects was possible. This experiment combines the strengths of the well established LIDAR technique[1] with stand-off Raman spectroscopy.[11]

Experimental

Pulsed stand-off Raman setup

The pulsed stand-off Raman system (Fig. 2) has been previously described in detail elsewhere[19] and only a brief description is given here. The setup comprised a Q-switched Nd:YAG NL301HT laser (EKSPLA, Lithuania) operating at 532 nm, with a pulse length of 4.4 ns and a 10 Hz repetition rate.

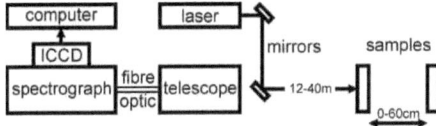

Fig. 2: Stand-off Raman setup for Raman LIDAR measurements

Raman scattered light was collected using a 6 Schmidt-Cassegrain telescope (Celestron, USA). The light was filtered through a 532 nm long pass filter (Semrock, USA) and collected via a fibre optical round-to-slit bundle cable consisting of 19 200-μm diameter optical fibres (Avantes, Netherlands). The bundle was directed to an Acton standard series SP-2750 imaging spectrograph. The light was detected by a PIMAX 1024RB intensified CCD (ICCD) camera (Princeton Instruments, Germany) with a minimal gate time of 500 ps. The ICCD camera was synchronised relative to the laser trigger so that the delay time between the excitation pulse leaving the laser and the Raman backscattered photons arriving at the ICCD could be adjusted. The laser pulse energy was set to power levels which avoided deterioration or destruction of the samples when the laser spot was 4 mm in diameter; 50 mJ for time-resolved Raman experiments and 20 mJ for range gated detection of $NaClO_3$ concealed in a white container at 40 m distance. Laser mirrors (Thorlabs, Germany) were used to align the laser coaxially to the telescope.

Time-resolved stand-off Raman spectroscopy

To gain spectral information at different distances the ICCD camera delay time was changed relative to the laser trigger. Therefore, only the Raman signal fraction from a certain distance could reach the detection system. The delay increases with the distance but also depends on the physical components of the setup, for instance on the length of the trigger cable between laser and camera. To obtain the time-resolved spectral information the delay time was changed in steps of 0.5 ns and the recorded spectra series was analysed using a Multivariate Curve Resolution-Alternating Least Squares algorithm (MCR-ALS).[20,21] This algorithm results in substance specific spectral data as well as in separate intensity profiles over time for each material. Based on these calculated profiles the delay time with the highest intensity was determined for each substance. For this a graphical user interface developed by Tauler et al.[22] implemented in MATLAB was used. The MCR-ALS algorithm relies on initial estimates as a starting point for the calculation.

Two methods exist to get initial estimates in order to obtain feasible results: when looking for known substances reference spectra can be used, whereas, if the nature of the analyte is unknown Evolving Factor Analysis (EFA)[20,21] generates initial spectra for the MCR algorithm. The advantageous combination of MCR-ALS and EFA does not require any information about the chemical composition of the sample, making this technique very interesting for stand-off Raman depth profiling of unknown samples.

To further improve the depth resolution, the obtained intensity profiles were fitted with the temporal laser pulse shape, which was previously determined via photodiode with a specified rise/fall time of 1 ns (DET10A, High Speed Silicon Detector, Thorlabs, Germany) and Oscilloscope (wave runner 64Xi, LeCroy).

For accurate position adjustment the samples were placed on a rail, which was coaxially aligned to the optical axis of the setup. The first sample was placed in a 10 mm-quartz cuvette 12 m away from laser and telescope. In order to quantify the depth resolution capability of the technique the distance between sample pairs was varied from 10 to 60 cm in steps of 5 cm.

The method accuracy was tested by assessing how well the determined distances correspond to the known distance between different substance pairs. Therefore, each two-sample-test system was validated and the standard deviation of the method was calculated. Furthermore, the method was extended to three different samples placed in a row to show, that the technique is not limited to a pair of substances only.

To investigate the NaClO$_3$ content in a white HDPE bottle at a stand-off distance of 40 m the camera delay time was varied from 315 to 340 ns in steps of 0.5 ns, whereas the gate time of 5 ns was kept constant. The sample was analysed accumulating 200 laser pulses with a pulse energy of 20 mJ and a wavelength of 532 nm.

Chemicals

Toluene (98%), o-xylene (98%), m-xylene (98%), p-xylene (99%) and NaClO$_3$ (>99%) all obtained from Sigma-Aldrich were used without further purification. A low density polyethylene sheet 1 mm in thickness (LDPE, Riblene FL 34 LDPE, Agru, Austria) was used as purchased. For the experiments the liquids were contained in quartz cuvettes 10 mm in length. A white chemical container made of high-density polyethylene (HDPE) (with a flat surface and wall thickness of 1.5 mm, dimensions 7.5 cm length and width with a height of 6 cm) was used to determine its NaClO$_3$ content.

One gram of NaClO$_3$ was compacted into a pellet using a pellet press. Therefore, the salt was ground in an agate mortar and compacted in three steps of increasing pressure (98, 157 and 235 bar), applied for 2 min per pressure step, producing pellets 1 cm in diameter with a thickness of 3 mm.

Confocal Raman microscope reference spectra

Reference Raman spectra were obtained using a confocal Raman microscope (LabRAM, Horiba Jobin-Yvon/Dilor, Lille, France) for comparison with the recorded stand-off Raman spectra as well as for the calculated spectral output from the MCR algorithm. Raman scattering was excited by a He–Ne laser at 632.8 nm and a laser power of 7 mW. The dispersive spectrometer was equipped with a grating of 600 lines/mm. The detector was a Peltier-cooled CCD detector (ISA, Edison, NJ, USA). The laser beam was focused manually on the sample by means of a ×20 microscope objective.

Results and Discussion

Stand-off Raman reference spectra

Stand-off Raman spectra of the used substances are shown in Fig. 3 for comparison, recorded at 12 m.

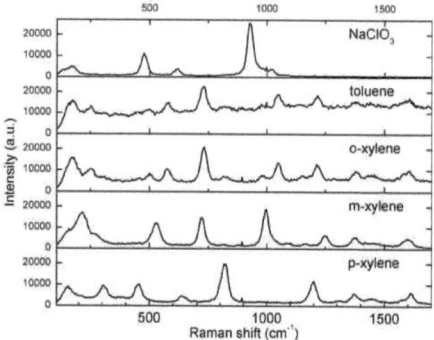

Fig. 3: 12 m-stand-off Raman reference spectra; gate width 5 ns, laser pulse energy 50 mJ

Determination of laser profile

The temporal intensity distribution of the 532 nm-laser pulse was determined via a photo diode. The average of 1000 laser pulses is shown in Fig. 4.

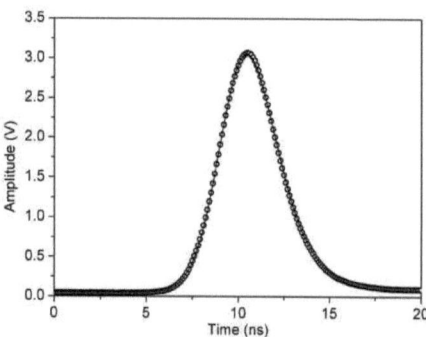

Fig. 4: Average 532 nm-laser pulse shape based on 1000 single pulses measured via photo diode; pulse length 4.4 ns defined by 1/e criterion; curve fit based on Equation 2

The average pulse shape (Fig. 4) can be described by a convolution of a Gauss function and an exponential function, detailed in Eq. 1, with an intensity correlation factor A, the time t, the standard deviation s, the exponential decay parameter λ and m which describes the position of the pulse in time.

$$I(t) = -A \cdot (0.2e)^{1/2} \cdot e^{\left(-t\lambda + m\lambda + \frac{s^2\lambda^2}{0.4e}\right)\left(erf\left(\frac{(\lambda s^2 - 0.2et + 0.2em)0.2e}{s}\right) - 1\right)}$$

Equation 1

Range-gated Raman spectroscopy

Toluene and p-xylene in 10 mm-quartz cuvette were placed at a stand-off distance of 12 and 12.6 m, respectively. The spectral change when shifting the ICCD camera delay from 133 to 146 ns is shown in Fig. 5.

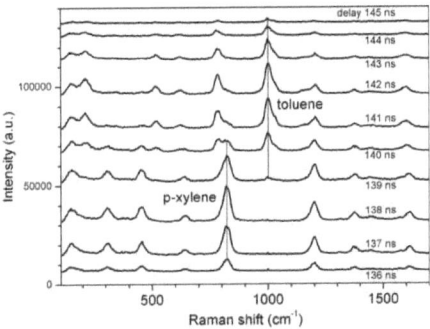

Fig. 5: Stand-off Raman spectra of p-xylene at 12.0 m and toluene at 12.6 m; different delay times in ns are stated next to the spectra

In Fig. 5 a selection of the recorded stand-off Raman spectra illustrate the spectral differences with increasing delay times between laser pulse and camera delay. Starting at lower delay times (bottom) the spectral features of p-xylene increase up to a maximum at a delay of 138 ns. Increasing the delay even more these p-xylene features decrease and the bands of toluene are detected, which is located 60 cm behind p-xylene.

For a more precise data analysis an MCR-ALS algorithm was employed to separate the spectral data set into intensity profiles over delay time for both substances (Fig. 6).

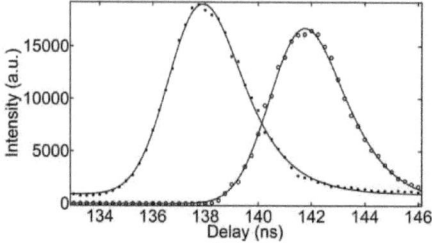

Fig. 6: Intensity profiles of p-xylene (black circles) and toluene calculated via MCR algorithm; curve fit based on exciting laser pulse shape (convolution of a Gauss and an exponential function)

In Fig. 6 the intensity profiles of p-xylene (134-140 ns) and toluene (140-146 ns) the maxima of the individual substance are well separated. To accurately determine these maxima without human intuition, a curve fit was employed based on the laser pulse shape, which was previously determined via photodiode (Fig. 4).

The time difference between the two intensity maxima was used to calculate the distance between the samples, using the speed of light. In addition to the intensity profiles over time the MCR-ALS algorithm offers spectral information as well, as can be seen in Fig. 7.

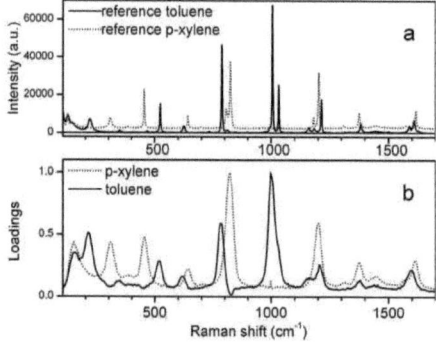

Fig. 7: Spectra of p-xylene and toluene; a: Raman reference spectra determined via confocal microscope; b: spectra resulting from MCR calculation

The spectra in Fig. 7 calculated by the MCR algorithm (b) compare very well to the reference spectra of the pure substances, recorded on a confocal Raman microscope (a).

Alternative algorithms

The MCR algorithm relies on a reasonable initial input to allow a successful calculation. The effectiveness of two alternative initial estimates was compared. For the first alternative measured reference spectra are used as a starting point for the iterative MCR algorithm. The second alternative calculates the initial estimates for the MCR algorithm without any prior knowledge of the samples, therefore facilitating an Evolving Factor Analysis (EFA).

To compare the accuracy of either method the distance between the two samples was varied between 10 and 60 cm in steps of 5 cm. For each of the 11 different sample arrangements the camera delay was changed, as described above, to obtain the depth resolved spectral data set. Then the MCR algorithm was applied for each configuration, using either measured reference Raman spectra or initial estimates calculated via EFA.

Fig. 8 shows a selection of range-gated stand-off Raman spectra for liquid o-xylene which was placed in front of a solid pellet of sodium chlorate ($NaClO_3$).

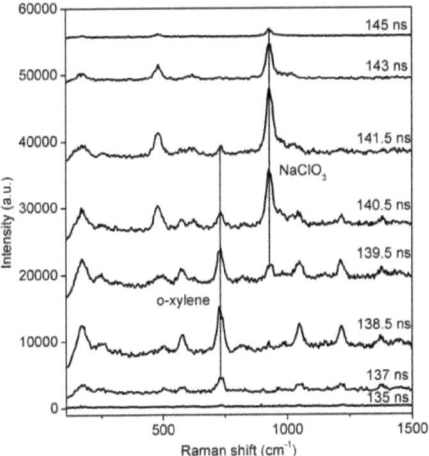

Fig. 8: Stand-off Raman spectra of o-xylene at 12.0 m and $NaClO_3$ at 12.45 m measured at different delay times (stated in the diagram)

Fig. 8 shows the rise of the spectral o-xylene features at lower delay times followed by high $NaClO_3$ intensities at higher delays, corresponding to the sample locations.

Based on the data sets recorded for 11 different distances between o-xylene and $NaClO_3$, the distances were calculated via MCR algorithm based on reference spectra or EFA initial estimates. To evaluate the accuracy of either algorithm, the calculated distances were compared to the real sample distances. To quantify the quality of both methods the calculated standard deviation of both methods is summarised in Table 1. The experiment was repeated using a different pair of analytes, namely toluene and low density ethylene (LDPE). The results of the latter experiment are stated in Table 1 as well.

Table 1: Accuracies for alternative MCR initial estimates for both sets of experiments

sample system	initial estimates	standard deviation of the method [cm]	relative standard deviation [%]
NaClO$_3$ + o-xylene	reference spectra	2.1	6.8
	EFA	5.4	18
LDPE + toluene	reference spectra	1.3	4.4
	EFA	1.9	6.4

Analysing toluene and LDPE it can be seen (Table 1) that the depth accuracy is similar for both compared initial estimates, ranging from 1.3 to 1.9 cm. However, for the sample pair NaClO$_3$/o-xylene the difference between the initial estimates is much more significant. The achieved accuracy is 5.4 cm when using the calculated EFA-estimates but improves to 2.05 cm when reference spectra are introduced to the MCR algorithm.

To prove that this measurement concept is not limited to two samples, three samples (m-, o- p-xylene) were placed at different distances from the setup (12.0, 12.2 and 12.4 m). According to the different sample distances the change in delay time led to substance specific intensity maxima in the spectra. For three xylene isomers (o-, m-, p-xylene) Fig. 9 illustrates the extension to more than two substances.

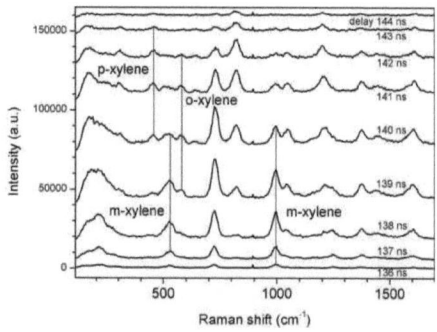

Fig. 9: Stand-off Raman spectra of m-xylene, o-xylene and p-xylene placed at 12.0, 12.2 and 12.4 m; measured at different delay times (stated in the diagram)

Using the MCR algorithm the delay times were calculated for each substance (Table 2). The time delay between each substance pair allowed the determination of the sample distances via the speed of light. An analogous experiment with toluene, o- and p-xylene was conducted, leading to similar results (Table 2).

Table 2: Calculation of relative sample distances d based on delay time differences t; the real distance between the samples was 20 cm

sample set 1	delay time (ns)	R^2	t (ns)	d (cm)
m-xylene at 12.0 m	138.4	0.987		
			1.6	23.9
o-xylene at 12.2 m	140.0	0.996		
			1.3	19.5
p-xylene at 12.4 m	141.3	0.978		

sample set 2	delay time (ns)	R^2	t (ns)	d (cm)
toluene at 12.0 m	138.0	0.995		
			1.7	25.5
o-xylene at 12.2 m	139.7	0.997		
			1.3	19.5
p-xylene at 12.4 m	141.0	0.978		

From Table 2 it can be seen that a depth accuracy of better than 5.5 cm was achieved for all cases. Two of the four distances were even determined with a deviation of only 0.5 cm.

Currently the limiting factor for the distance accuracy is the laser pulse length of 4.4 ns, which corresponds to a length of 1.32 m. Even if the application of chemometric algorithms allows to improve the accuracy of the technique, using a ps-laser is essential the fully exploit the potential of the method, as shown for close sample distances by Iping Petterson et al.[18]

Time-resolved stand-off Raman for determination of container content

At a distance of 40 m, the content of a flat bottle (white HDPE, wall thickness 1.5 mm) containing solid NaClO$_3$ was examined using varying camera delays. At close proximity, this principle was already applied to powder layers[17] as well as to substances behind polymers.[18,23] In this work we extend the measurement range to 40 m allowing the detection of potentially dangerous substances concealed in containers which are non-transparent for the human eye. By changing the camera delay time this technique offers an alternative to stand-off SORS (spatial offset Raman spectroscopy)[24], a method which relies on diffuse scattering of the excitation laser in the sample under investigation.[25]

In Fig. 10 spectra of solid NaClO$_3$ in a white HDPE container are shown. By changing the camera delay time it is possible to distinguish between the plastic container (a) and the NaClO$_3$ content (b).

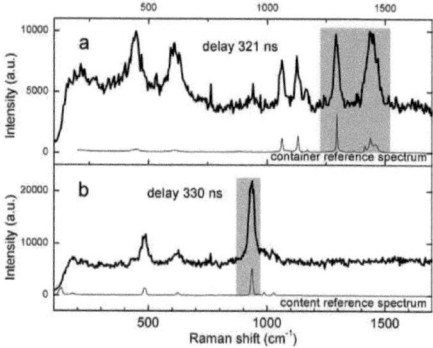

Fig. 10: Influence of camera delay time on 40 m-stand-off Raman spectra of solid NaClO$_3$ content in a white HDPE container with a wall thickness of 1.5 mm; a: delay of 321 ns shows HDPE container spectrum; b: delay of 330 ns reveals spectrum of NaClO$_3$ content; grey boxes indicate range for baseline corrected integration of substance bands

In Fig. 10 the spectral ranges for baseline corrected integration are indicated by the grey boxes; 1220-1520 cm^{-1} for HDPE and 870-970 cm^{-1} for NaClO$_3$ (reference spectrum in Fig. 3).

The change of these integrals with varying delay time is shown in Fig. 11.

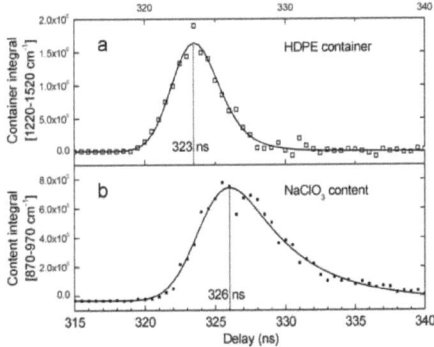

Fig. 11: Change of band intensities with delay time; a: baseline corrected integral (1220-1520 cm^{-1}) of HDPE container band; b: baseline corrected integral (870-970 cm^{-1}) of NaClO$_3$ content band

In Fig. 11 the maximum of the container integral (a) is detected at a delay time of 323 ns whereas the most intense content signal (b) is recorded 3 ns later. A delay difference of 3 ns corresponds to a sample distance of 45 cm, which is longer than the dimensions of the sample container. Longer than expected photon migration times have been observed in other experiments.[18,23] It can be explained by the fact that photons have longer net migration times in the container material (HDPE) and in the solid NaClO$_3$ relative to air due to the scattering properties, refractive indices and thicknesses of the materials. In fact, the net migration times of photons in different polymers have been measured[23] and it has been determined that a Raman signal is delayed approximately one order of magnitude lower than the speed of light in a transparent organic material (2.0 x 10^8 m/s). A rough "back-of-the-envelope" calculation can be made regarding the delay between the photons detected from the container and those from the middle of the sample (NaClO$_3$) using some general assumptions. Assuming that the sample container and content are non-absorbing, it can be postulated that the photon migration time in the polymer container and the solid sample NaClO$_3$ is roughly the same and that the maximum signal from NaClO$_3$ will occur in the middle of the container. The total direct distance from the surface of the container to the middle and back was measured to be 7.5 cm. Thus using a photon migration time for polythene[23] of 2.0 x 10^7 m/s, the expected delay would be roughly 4 ns consistent with the experimental results obtained. Fig. 10 and Fig. 11 show that using appropriate camera delay times it is possible to specifically investigate the content in a container at a distance of 40 m, which is non transparent to the human eye.

Conclusion

Using the speed of light it is possible to determine the position of samples. A potential application could be the detection of a non-fluorescing sample located proximally in front of a fluorescing container since the time scale on which Raman photons are generated is usually much shorter compared to fluorescence.

It has also been demonstrated that using a MCR algorithm, range-gated stand-off Raman spectroscopy adds molecular specific information without any prior knowledge of the investigated sample material. Applying a multivariate curve resolution algorithm it was demonstrate that using a pulsed laser and a short gated ICCD camera it is possible to distinguish between Raman signals originating from different samples positioned at different stand-off distances. A depth resolution of typically within 7% of the measured distance for this technique was obtained. In addition three sample systems were analysed to show the flexibility of this technique due to high amount of chemical information, that Raman scattering offers. Furthermore, $NaClO_3$ was identified in a white plastic container at a distance of 40 m, showing that stand-off Raman spectroscopy is capable of detecting substances in containers which are non-transparent to the human eye.

Acknowledgement

The research leading to these results has received funding from the European Community's Seventh Framework Programme (FP7/2007-2013) under Grant Agreement No 218037. Further financial support was provided from the Austrian Science Foundation (RSA project no. 818664).

Literature

1. A. Wehr. "Airborne laser scanning—an introduction and overview". ISPRS J. Photogramm. Remote Sens. 1999. 54(2-3):68-82.
2. M. Laurenzis, F.Christnacher, E. Bacher, N. Metzger, S. Schertzer, T. Scholz. "3D range-gated imaging in scattering environments". Proc. of SPIE. 2011. 8186A:03-13.
3. J. Reichardt, S. E. Bisson, S. Reichardt, C. Weitkamp, B. Neidhart. "Rotational vibrational-rotational Raman differential absorption lidar for atmospheric ozone measurements: methodology and experiment". Appl. Opt. 2000. 39(33): 6072-9.
4. F. de Tomasi, M. R. Perrone, M. L. Protopapa. "Monitoring O_3 with solar-blind Raman lidars". Appl. Opt. 2001. 40(9):1314-20.
5. M. Wu, M. Ray, K. H. Fung, M. W. Ruckman, D. Harder, A. J. Sedlacek. "Stand-off Detection of Chemicals by UV Raman Spectroscopy". Appl. Spectrosc. 2000. 54(6): 800-806.
6. M. D. Ray, A. J. Sedlacek, M. Wu. "Ultraviolet mini-Raman lidar for stand-off, in situ identification of chemical surface contaminants". Rev. Sci. Instrum. 2000. 71(9):3485-3489.
7. S. K. Sharma, S. M. Angel, M. Ghosh, H. W. Hubble, P. G. Lucey. "Remote Pulsed Laser Raman Spectroscopy System for Mineral Analysis on Planetary Surfaces to 66 Meters". Appl. Spectrosc. 2002, 56(6):699-705.
8. J. C. Carter, S. M. Angel, M. Lawrence-Snyder, J. Scaffidi, R. E. Whipple, J. G. Reynolds. "Standoff detection of high explosive materials at 50 meters in ambient light conditions using a small Raman instrument". Appl. Spectrosc. 2005. 59(6): 769-75.
9. S. K. Sharma, A. K. Misra, B. Sharma. "Portable remote Raman system for monitoring hydrocarbon, gas hydrates and explosives in the environment". Spectrochim. Acta, Part A. 2005. 61(10):2404-12.
10. S. K. Sharma, A K. Misra, P. G. Lucey, S. M. Angel, C. P. McKay. "Remote pulsed Raman spectroscopy of inorganic and organic materials to a radial distance of 100 meters". Appl. Spectrosc. 2006. 60(8):871-6.
11. A. Pettersson, I. Johansson, S. Wallin, M. Nordberg, H. Östmark. "Near Real-Time Standoff Detection of Explosives in a Realistic Outdoor Environment at 55 m Distance". Propellants, Explos., Pyrotech. 2009. 34(4):297-306.
12. M. Akeson, M. Nordberg, A. Ehlerding, L.-E. Nilsson, H. Ostmark, P. Strömbeck. "Picosecond laser pulses improves sensitivity in standoff explosive detection". Proc. of SPIE, 2011. 8017: 80171C-80171C-8.
13. Y. Fleger, L. Nagli, M. Gaft, M. Rosenbluh. "Narrow gated Raman and luminescence of explosives". J. Lumin. 2009. 129(9): 979-983.
14. J. Moros, J.A. Lorenzo, P. Lucena, L. M. Tobaria, J.J. Laserna. "Simultaneous Raman spectroscopy-laser-induced breakdown spectroscopy for instant standoff analysis of explosives using a mobile integrated sensor platform". Anal. Chem., 2010. 82(4):1389-1400.

15. J. Moros, J.A. Lorenzo J.J. Laserna, "Standoff detection of explosives: critical comparison for ensuing options on Raman spectroscopy-LIBS sensor fusion". Anal. Bioanal. Chem. 2011. 400(10):3353–3365.
16. R. M. Measures. "Fundamentals of Laser Remote Sensing". In: R. M. Measures, editor. Laser Remote Chemical Analysis. NY: John Wiley & Sons, 1988. Pp. 1-83.
17. P. Matousek, N. Everall, M. Towrie, A W. Parker. "Depth profiling in diffusely scattering media using Raman spectroscopy and picosecond Kerr gating". Appl. Spectrosc. 2005. 59(2):200-205.
18. I. E. Iping Petterson, M. López-López, C. García-Ruiz, C. Gooijer, J. B. Buijs, F. Ariese. "Noninvasive Detection of Concealed Explosives: Depth Profiling through Opaque Plastics by Time-Resolved Raman Spectroscopy". Anal. Chem.2011. 83(22): 8517-8523.
19. B. Zachhuber, G. Ramer, A. Hobro, E. t. H. Chrysostom, B. Lendl. "Stand-off Raman spectroscopy: a powerful technique for qualitative and quantitative analysis of inorganic and organic compounds including explosives". Anal. Bioanal. Chem. 2011. 400(8):2439-2447.
20. R. Tauler, "Multivariate curve resolution applied to second order data". Chemo. Intell. Lab. Syst. 1995. 30(1):133-146.
21. R. Tauler, A. Izquierdo-Ridorsa, R. Gargallo, E. Casassas. "Application of a new multivariate curve resolution procedure to the simultaneous analysis of several spectroscopic titrations of the copper (II) -polyinosinic acid system". Chemo. Intell. Lab. Syst. 1995. 27:163-174.
22. J. Jaumot, R. Gargallo, A. de Juan, R. Tauler. "A graphical user-friendly interface for MCR-ALS: a new tool for multivariate curve resolution in MATLAB". Chemo. Intell. Lab. Syst. 2005. 76:101-110.
23. I. E. Iping Petterson, P. Dvořák, J. B. Buijs, C. Gooijer, F. Ariese. "Time-resolved spatially offset Raman spectroscopy for depth analysis of diffusely scattering layers". Analyst. 2010. 135(12):3255-3259.
24. B. Zachhuber, C. Gasser, E.t.H Chrysostom, B. Lendl. "Stand-off spatial offset Raman spectroscopy for the detection of concealed content in distant objects". Anal.Chem. 2011. 83(24):9438-9442.
25. P. Matousek, I.P. Clark, E.R.C. Draper, M.D. Morris, A.E. Goodship, N. Everall, M. Towrie, W.F. Finney, A.W.Parker. "Subsurface probing in diffusely scattering media using spatially offset Raman spectroscopy". Appl. Spectrosc. 2005. 59(4):393-400.

i want morebooks!

Buy your books fast and straightforward online - at one of world's fastest growing online book stores! Environmentally sound due to Print-on-Demand technologies.

Buy your books online at
www.get-morebooks.com

Kaufen Sie Ihre Bücher schnell und unkompliziert online – auf einer der am schnellsten wachsenden Buchhandelsplattformen weltweit! Dank Print-On-Demand umwelt- und ressourcenschonend produziert.

Bücher schneller online kaufen
www.morebooks.de

VDM Verlagsservicegesellschaft mbH
Heinrich-Böcking-Str. 6-8   Telefon: +49 681 3720 174   info@vdm-vsg.de
D - 66121 Saarbrücken   Telefax: +49 681 3720 1749   www.vdm-vsg.de

Printed by Books on Demand GmbH, Norderstedt / Germany